THE NEUTRINO CASE

Marcello Veneziano

Publications by same author:

Black Holes And The Secret of Life	2009
Travel into a Small Universe	2007
EMI- RFI Filtering	1992

" Made we weren't to live like brutes,
but to pursue virtue and knowledge".

Ulysses in Dante's "The Divine Comedy"

To Giorgio and Lara with love,
Wishing that their generation
And the ones to follow will
Continue this journey of virtue
And knowledge started ages ago.

Contents

INTRODUCTION

Ten years ago I began writing about a nuclear particle called the "neutrino" because I started developing a new theory concerning the neutrino behaving like a small black hole. Back then, this nuclear particle had not much prominence in the world of nuclear physics. Today, due to new scientific developments, this particle is one of the most studied. There is a belief that these new studies could lead to the discovery of some of the most important secrets of the Universe. The problem with the neutrinos is that they cannot be observed in the very high energy particle accelerators. Basically no trace of them was ever found during the numerous experiments in many of these facilities. There are only a few places in the world where they can be observed interacting with matter inside specially made large tanks. These tanks are located thousands of feet underground so that the neutrino interactions observed are not mistaken for cosmic rays. Unfortunately, only a few neutrino interactions with the surrounding matter are found even in these locations that are designed for their detection. Neutrinos are so small and interact so little, that I believe there is a low expectation of ever discovering the secrets of their nature by direct observation. I think that the best way to explore the neutrino is to observe the entire Universe and try to find a logical explanation for the role they play in it. This is exactly what I am about to attempt in this book.

I believe that neutrinos are small black holes and that some of these neutrinos create other nuclear particles by capturing gamma rays in their event horizons. The neutrinos are located in the center of these newly created particles, just like large black holes are located in the center of the galaxies. Since "no smoking gun" may ever be found to prove my theory, I will try to put together many cases of circumstantial evidence. To use an example, a man is on trial for murder and there is only one piece of circumstantial evidence against him. If this evidence has only a 90% probability of being correct and he gets convicted, then there is a 10% chance that the man sent to jail is innocent. But if the same man had 10 pieces of circumstantial evidence against him, and all having the same 90% probability to be accurate, then if the man is convicted the probability of his innocence would be only one in a billion. In this case he could be convicted by the jury with little doubt. After all, the inmate population of the entire world is far less than one billion individuals and if every one of them had been convicted with the same probability of innocence stated directly above, very likely there would be no innocent people in jail. We know that unfortunately this is not the case.

The above analogy was used because I believe that if this book succeeds in presenting many cases of circumstantial evidence in which the neutrino appears to behave as a black hole, then the reader will probably believe that this representation of the neutrino is correct. In this book twenty cases of circumstantial evidence will be introduced. Of course it is impossible to assign to each case a probability of its correctness since this probability assignment is very subjective. However, the reader should keep in mind that if each case had a 50% probability to be correct, then the theory presented would have only one chance in a million to be wrong in all twenty cases. If half of the cases are discarded because of low credibility, then the theory would have only one chance in a thousand to be incorrect. A 99.9% probability that this theory is correct still provides a very convincing argument.

Since a new particle model is introduced in order to give the reader a good perspective, I will provide some historical facts on the development of the present nuclear particle model. Hopefully this will help to clarify the circumstances that brought nuclear physics to embrace the present Standard Model. The reader needs to know that the quarks theory is based on some circumstantial evidence too. This theory, when it was originated in the year 1963, could have generated considerable suspicion, since it was started as a method to catalog the more than 100 stable and unstable particles already known. At that time, Murray Gell-Mann, the scientist credited for the theory, called the quark a mathematical entity instead of a real entity, probably thinking that the existence of quarks would never be proven. Around 1970, Dick Taylor, Henry Kendall and Jerry Friedman using high energy electrons were able to look inside the structure of protons and neutrons. Since they were unable to find any solid masses, they concluded that any component inside these particles had to be very small and similar to the point like ½-spin structure of the electron and neutrino. The findings that got the most attention were generated when they saw fractions of the electron charge in their experiments. These findings created much enthusiasm in the scientific community because they were assumed to be the proof of the validity of the quarks theory which assigns to quarks 1/3 or 2/3 electron electrical charges. But the quarks structure formulated for the protons and neutrons is not the only possible structure that would match the results recorded by the Taylor, Kendall and Friedman team. As matter of fact, I'm confident that the new particle model introduced in this book satisfies the two major findings of the Taylor, Kendall and Friedman experiment. First, it satisfies the point like ½-spin requirement because it has a point like ½-spin neutrino in the center of the structure. Second, this new particle model would show fractions of electron charges when analyzed using high energy electron beams.

 Replacing the quarks model with another model would be almost a futile exercise unless the new model shows considerable advantages. But this is exactly our case since this model is able to explain the workings of the gravitational field. The incapability of the present Standard Model to explain gravitation is one of its major failings. Therefore gravitation is the main reason why the acceptance of this new model theory would be greatly desirable. The new model, because it explains the intricacy of the gravitational force, allows nuclear physics to connect to astrophysics. The joining of these two sciences will produce a major step forward in the discovery of the true nature of the universe. This is the major reason why scientists should give this presentation a thorough evaluation before making a decision on its merits.

The quarks theory was developed almost 50 years ago. This presents a major obstacle to the introduction of new theories because, now, there are two generations of physicists that have been educated with the idea that only the quarks theory is acceptable. For a long time this theory has been assumed to be correct without using any rigorous proof in the process of its study. Unfortunately we know that anything assumed to be an undisputable truth can be very hard to be proven incorrect. One can sense an almost religious approach to the quark theory, when no skepticism arises in the scientific community, after experiments with high-energy-electro-electron collisions (non quarks particles) show the same byproducts of proton-proton collisions. Established ideas are a hurdle very difficult to overcome and that is one of the reasons I opted for the format of a trial to introduce what I call circumstantial evidences of the neutrino-black-hole case. I hope that this format is the best for the task ahead since I know the tremendous amount of skepticism that this new theory will have to face.

The first two chapters, besides talking about the nuclear particles in general, cover the basis of our theory. In the third chapter we start with the NEUTRINO CASE No. 1. Each of the following chapters will cover a new CASE in which we show why we believe the neutrino behaves like a black hole.

Chapter I

The Nuclear Particle

In this Chapter, we explain how the neutrinos can create the other nuclear particles. To simplify our task, we subdivide the nuclear particles into two major subgroups: stable and unstable particles. Unstable particles have a short life span. In the last fifty years, many unstable particles have been discovered that have life spans of less than microseconds. In this book we are concerned only with stable particles, and there are only three stable particles in the Universe: the *electron*, the *proton* and the *neutrino*. The *neutron* is not included because, having a decay time of a few hundred seconds, it cannot be considered a stable particle. The neutrino is by far the smallest of the stable particles. Its mass is millions of times smaller than the mass of a proton, and it has an interaction rate so low that it can cross the Earth's diameter without colliding with any Earth matter.

During the next few chapters, we will come to the conclusion that, indeed, there exists only one particle in the Universe, and this particle is the *neutrino*. The other particles are a combination of neutrinos and photons; these particles are formed when a photon is captured by the gravity field of a neutrino acting as a black hole. The reader will probably be skeptical that such a small particle can behave like a black hole, but a black hole is defined by the density of its matter, not just the quantity of it, and, in our theory, the neutrinos are by far the most dense matter in the Universe.

Thousands of years after the Big Bang the photons started appearing (as will be explained in the next chapters).The neutrinos started capturing the new developed photons. The concept of a particle capturing a photon is not new. In the photoelectric effect, electrons capture photons with wavelengths that are millions of times bigger than the electrons' dimensions. The basic difference between our theory and the photoelectric effect is that the photons captured in our theory have much shorter wavelengths. Three different types of neutrinos are presently known: the *electron neutrino*, the *muon neutrino*, and the *tau neutrino*. The Oscillating Neutrino Theory,[1] which we embrace in this book, tells us that the electron, the muon, and the tau neutrino are just different phases of the same particle and, therefore, we chose to call them all with the general name: *neutrino*. The reader does not need to know more about this theory since the phenomena is not clear to the scientist either.

The photons that we mention in developing our theory are just very high-energy **electromagnetic waves (gamma rays)**. **Electromagnetic waves** cover a wide spectrum of energy that goes from radio waves to gamma rays; their energy spectrum varies from a fraction of an electron volt to billions of electron volts. The energy of these waves can be represented by multiplying the frequency of the wave by a constant (Planck constant). Therefore, the frequency is the only variable that defines the different electromagnetic waves. You may ask why a new particle model is needed. The truth is that, after almost forty years from the development of the present nuclear model (Standard Model), there are still common well spread doubts on its validity. We can read, on the 2010 version of the Internet encyclopedia Wikipedia, the following excerpt concerning the Standard Model:

Why are there three <u>generations</u> of particles in the Standard Model? Is the Standard Model a good approximation to reality, or is it fatally flawed? The Standard Model falls short of being a <u>complete theory of fundamental interactions</u>, primarily because of its lack of inclusion of <u>gravity</u>, the fourth known fundamental force.

There have not been any major discoveries in nuclear particle physics for many years. This lack of progress can be attributed to the methodology of the research. One of the problems has been the almost exclusive use of high-energy accelerators for colliding particles. These high-energy accelerators, by colliding particles close to the speed of light, create an artificial environment that is totally different from our observable Universe. In this book, we will observe the stable nuclear particles with the same methods that we observe the solar systems, where gravitation and the General Theory of Relativity are used to explain most of the phenomena. The reason that gravitation is not believed to play a major role in the workings of the inner structure of the particles is that gravitation is a considerably weak force when compared to the other forces that work at the atomic level. This assumption is no longer valid when we address the gravitational field of the black holes inside their event horizon [2]. The black hole gravitational field is the only force capable of capturing every kind of matter, including every type of electromagnetic wave, from light rays to gamma rays. What is presented in this book is not just a theory; the nuclear particle model that we develop will give us the capability to compute the magnetic fields of the electron and muon with great accuracy when compared to laboratory data. Furthermore, it gives us a new understanding of the concept of the electrical charge. The neutrino is the main protagonists of this book; a mysterious particle whose existence was only surmised until a few years ago. Its existence was proved when its interactions with matter were finally observed. This particle is elusive because of its very small mass and its lack of any electrical charge (as the name implies).

In the next few chapters, first we'll develop a new particle model, and then we'll prove the credibility of the new model by testing it against existing scientific data and facts.

Footnotes:

1.   The Oscillating Neutrino Theory is based on the idea that the neutrino mass can vary (oscillate) while traveling in space.

2. The point in space of no return for a black hole. Not even light can escape once it crosses this point, therefore the name black hole.

Chapter II

The Black Hole Model
Space and
Time Warping

In this chapter we try to define what kind of black hole the neutrino would be. We are going to explain how small black holes and large black holes have many similarities and a few differences.

Small black holes can be represented by a model that is exactly the same as the model of the gigantic black holes, with many times the mass of the Sun that we find throughout the Universe. This model was introduced the first time in 1916 by the German astronomer Carl Schwarzschild at a time when the existence of black holes was only a hypothesis. Nowadays their existence is widely accepted and so is the presence of a super massive black hole in the middle of every galaxy. Each of these super massive black holes has a mass many millions the solar mass. (Solar mass is the size of our Sun and is used as a measurement to express mass in astronomy).

Large Black Holes

The Schwarzschild black hole model is depicted in *Figure 2.1*. This black hole model consists of three components: the singularity, the event horizon, and the photon sphere. Located in the center of the black hole is the singularity.

This mathematical name is used because, in this region, the space and time fabric of the Universe develops a great discontinuity. The same thing happens in certain math functions. The **singularity** in large black holes consists of a matter as dense as the nucleus of an atom. When a body achieves this density, the amount of mass needed for it to become a black hole is equivalent to a few solar masses. If the Earth were to be compressed to the same density of the nucleus of an atom, it would be reduced to the size of a large boulder, but still it would not reach a singularity status. Therefore, both mass and density requirements have to be met in order for a singularity to be formed.

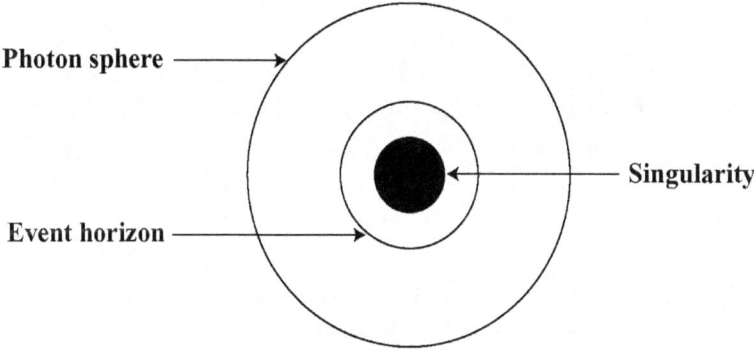

Figure 2.1 - The Black Hole Model

In *Figure 2.1*, the singularity is represented by the black circle in the center. Around the singularity is the event horizon that includes that portion of space where nothing—not even light—can escape singularity's gravitational pull. Beyond the event horizon, is the photon sphere. The photons in this space do not precipitate to the center of the singularity, but are forced to stay forever in an orbit surrounding it. Photons normally travel in a straight line (the way the Sun's rays travel to Earth), but, in the photon sphere, they travel in a circular orbit.

How is the existence of a very small mass black hole possible, when, for example, the Earth has not enough mass to become a black hole? The answer to this question is that large black holes cannot achieve a density greater than the density of the nucleus of an atom, while the small black holes do not have such limitation. In facts, small black holes singularities have to be trillions and trillions of times denser than the nucleus of an atom to achieve their properties.

Small Black Holes

For a small black hole to exist, its singularity must consist of a matter much denser than the nucleus of an atom. This means that nuclei particles like the proton and the neutron do not qualify. Of the remaining particles—the electron and the neutrino—only the neutrino is a possible candidate to become the singularity in a small black hole. In the next few chapters, the reason for this will become evident. We can still use the Schwarzchild model in *Figure 2.1* to represent our small black hole. Inside the singularity, instead of a mass a few times greater than the solar mass, we place the neutrino, a particle whose mass is millions of times smaller than a proton mass. We have to remind ourselves that the radius of the singularity is more relevant than the amount of its mass. The radius defines the actual density (specific weight) of the singularity. The density creates the black hole effect, and, according to our theory, there is nothing denser than a neutrino.

[20]

A particle generates a gravitational field **G** on its surface. That gravitational field is directly proportional to the mass and inversely proportional to the radius.

G is proportional to M/r²

Where **M** is the mass of the particle, and **r** is the radius of the particle. Therefore, a mass with a very small radius can generate a very powerful gravitational field in its immediate proximity.[1]

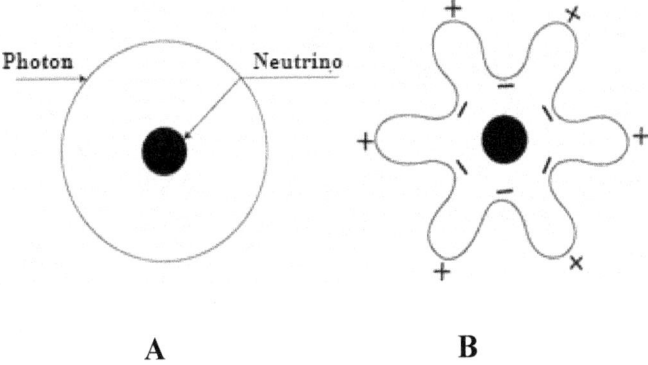

A   **B**

Figure 2.2- The Polarized Photon and its Standing Wave Electrical Field

The event horizon of a neutrino is extremely small—many orders of magnitude smaller than the size of a proton. The photon sphere (*Figure 2.1*) is where the neutrinos are different from large black holes. Neutrinos, to form other stable particles like the electron and the proton, can have only a single photon surrounding them. The wave frequency or energy of this single photon is determined by the particle mass. The wave frequency of this single photon can be derived using Einstein's famous equation, in which the mass of a particle multiplied by the square of the speed of light is equal to the total energy of the particle:

$$E = M_{particle} \, C^2$$

We use this equation by saying that the photon carries most of the particle energy. The small amount of energy of the singularity mass is temporarily neglected. Then, the same energy equivalence can be obtained using the photon wave frequency υ and the Planck constant h.

$$E = M_{particle} \, C^2 = h\upsilon$$

In the next chapters we'll use this equality frequently.
In the following paragraphs I will introduce a concept that is fundamental to the development of our theory.

In *Figure 2.2*, both illustrations, **A** and **B**, are of exactly the same particle. Illustration **A** depicts the particle with a single polarized photon surrounding a singularity or neutrino. Illustration **B** depicts the same particle and shows only the electrical field of the photon. The beginning and the end of the electric field are in phase with each other and form a **standing wave**. What we call a photon is essentially a short-duration electrical field and a magnetic field, orthogonal to each other, moving in space at the speed of light. For simplicity, *Figure 2.2 B* depicts the outside of the wave as positive and the inside as negative; the opposite (with the outside negative and the inside positive) is just as valid. We selected in this case a six-cycle wave, as you can verify by counting the peaks or the valleys. No matter how hard we try, it is impossible to draw the outside half of the wave in the same shape as the inside half wave. Even the areas included inside each semi-wave are different in size. There is an explanation for these anomalies. For the wave to curve into a circle, the travel of the inner semi-waves has to be reduced by a total of 360 degrees. These are the actual degrees needed for the photon to move in a circle.

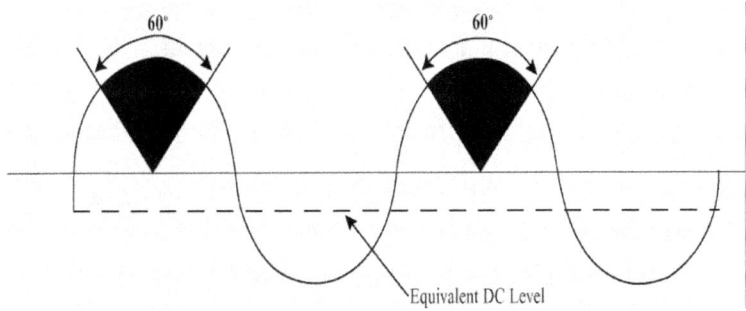

Figure 2.3 - Two Cycles of the Six-Cycle Standing Wave

Figure 2.3 depicts two cycles of the six-cycle wave that is shown in *Figure 2.2 B*. The neutrino is assumed to be located on top of *Figure 2.3*. The shaded areas depict the 60 degrees that are missing for each of the six cycles. When we add together the missing degrees of the six cycles (60 degrees each), we get 360, or a complete revolution. The missing degrees bend the photon path and enable the photon to orbit the neutrino. We can deduce that, in order for a neutrino to capture a photon, it has to be able to slow down the travel of the inner wave.

We know that travel distance can be found by multiplying speed by the amount of travel time. In our case, speed cannot change because the photon is moving at the speed of light; therefore, the only way to reduce the amount of travel is to slow down time. Einstein's General Theory of

[23]

Relativity tells us that the gravity field affects the time clock. In the case of a particle, the stronger the gravity field, the slower the particle's internal clock.[2] This is the solution to our problem. The inner part of the wave—the one that grazes the neutrino event horizon—experiences a stronger gravity field and, therefore, a slower clock. The inner wave, not being able to travel as far as the outer part of the wave, forces the photon into a circular orbit.

A photon traveling in free space has a positive semi-wave identical to the negative semi-wave. Both semi-waves cancel each other out with no trace of a positive or negative electrical charge. In *Figure 2.3* we show that, in the case of the captured photon, because of the 60 degrees missing in the negative semi-wave, there is an imbalance between the two semi-waves. This imbalance creates a charge because the positive semi-wave is not totally counterbalanced by the negative semi-wave. In Chapter X, we will show that the charge of a captured photon is always the same no matter its frequency.

To be stable, small black holes can have only one orbiting polarized photon. The end and the beginning of the photon connect to each other, making a standing electromagnetic wave. Only two of these configurations are stable—the electron and the proton. There is one concept we must remember when we think of the differences between small and large black holes: their masses do not cover a continuous spectrum. The singularities of large black holes must have masses of a few solar masses or more.
  The singularity of a small black hole can have only one mass — the mass of the neutrino. Small black holes must be ½ spin particles otherwise they would easily aggregate

with each other becoming quickly large black holes. This concept will be clarified in the next few chapters.

Some physicists believe that small black holes, even if their existence could be possible, would quickly evaporate. This evaporation would result from the effect of the formation of particle pairs at the black hole event horizon. The neutrinos are orders of magnitude too small for this effect to become possible, therefore we do not consider this to be a problem. The neutrinos do not quickly evaporate but, under certain conditions, they can decay.

Figure 2.3 shows an *equivalent DC*[(3)] *level* because there is an imbalance between the two semi-waves electromagnetic fields. This equivalent DC level would have an undetectable ripple because of the extremely high frequency; therefore, we decided to show it as a straight line.

1. In Chapter XIV, there is a more detailed mathematical presentation of the black hole.

2. Particles have their own internal clock, which is not absolute but relative. This internal clock experiences changes when the particle is exposed to a potential energy field. The particle's own photon, with its frequency, defines its clock. The neutrino has no photon; therefore, it has no clock. This is consistent with the General Theory of Relativity idea that, inside a singularity event horizon, time stops elapsing.

3. DC stands for Direct Current. In our case it stands for an electrical charge derived from a warped electromagnetic field.

Chapter III

The Planck Length

As we explained in the introduction, this book is all about presenting different cases in which the neutrino can be construed to have the nature of an infinitesimally small black hole. The case that we present in this chapter is important because it demonstrates that a black hole neutrino fits in a pre Big Bang Universe and is coherent with the Big Bang scenario accepted by most astrophysicists.

NEUTRINO CASE No. 1

Most of the times after an explosion, it is possible to analyze the remains and generate an approximate picture of the way things were before the explosion. We think that the Planck length could be one of those remains that give clues about the universe before the Big Bang. The Planck length is equal to $(h \ G/c^3)^{1/2} = 1.6 \cdot 10^{-35}$ meters and includes most of the constants that are significant in our universe. The G stands for the universal gravitational constant, the h for the Planck constant and the C for the speed of light. We proceed with the following analysis hoping that it will lead us to one of the clues that we are searching for. In this analysis we assume that the universe had the radius of the Planck length at the time of the Big Bang and therefore these universal constants became a permanent part of the expanding universe. Most astrophysicists agree that the universe was much smaller than an atom then. Though the Plank length is many orders of magnitude smaller than an atom, this is still a possible assumption.

Astrophysicists also agree that, at the time of the big Bang, the first particles created were the neutrinos. We assume that neutrinos were part of the universe even before the Big Bang since in such a small universe no other nuclear particle would have been able to fit inside it. This early universe was about 10^{20} times smaller than a proton. Now, we proceed with the following computations. If the universe at the time of the Big Bang had a radius equal to the Planck length the universe volume would have been 10^{-104} m^3. At this time, we make another assumption. We assume that the neutrino is a black hole, and we compute the neutrino event horizon by using a neutrino mass of 1.0 ev/c^2. With this mass, the event horizon of the neutrino is 10^{-63} m in radius. We compute the volume of the neutrino event horizon to be 10^{-189} m^3. If we divide the volume of a universe with a Planck radius, by the neutrino event horizon volume, we compute that at least 10^{85} neutrinos could have fitted inside this pre Big Bang universe. By converting 10^{85} neutrinos into mass, we get 10^{50} kg. This figure is close to the 10^{52} kg of mass which is considered to be the approximate mass of the present universe. If we consider that neutrinos heavier than the electron neutrino could have been part of this mix, then we could say that a pre Big Bang universe with only neutrinos could easily match the mass of the present universe. Therefore, if the universe was ever as small as the Planck length, it is very likely that neutrinos were part of it, and in order to fit they would have to have a density of about 10^{155} kg/m^3. With such a density these neutrinos would have been black holes as it is computed in Chapter XIV.

[27]

In our analysis the neutrino as a black hole and the pre Big Bang universe share the same density of 10^{155}kg/m^3. They also share the characteristic of oscillating. Can this be the case, that matter, when it is so dense, becomes unstable? One possibility is that the pre Big Bang universe, being filled with oscillating particles, would also be forced to oscillate. In this case, when all the 10^{85} oscillating neutrinos got in phase with each other, it is predictable that there would have been the great explosion that we call the BIG BANG.

From this analysis we were able to show that it was possible for a pre Big Bang universe with a radius as small as the Planck length to contain all the matter of the present universe. Also we were able to show how oscillating black hole neutrinos could have caused the Big Bang.

Chapter IV
The Magnetic Dipole

NEUTRINO CASE No. 2

The fact that there is a magnetic field associated with nuclear particles has been known for decades. The enigma about the nuclear particle magnetic field is that smaller particles have a stronger magnetic field than larger particles. This is the equivalent of a 100 pound man lifting twice as much as 200 pound man. To solve this enigma we have to analyze why particles have a magnetic field. One of the basic facts is that, whenever there is an electric charge or current rotating in a circle, a magnetic field is generated. The magnetic field in this case is more properly called a *magnetic dipole moment* as shown in *Figure 4.1.*

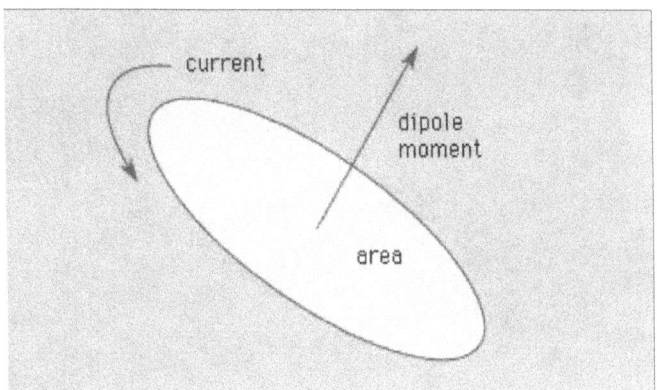

Figure 4.1- Magnetic Dipole Moment
This dipole acts exactly as a small magnet and aligns itself with any exterior magnetic field, just as the needle of a compass aligns with the Earth's magnetic field.

The magnitude of the dipole is computed by multiplying the current by the area.

In the case of the nuclear particle, we compute the equivalent current by multiplying the particle charge by the frequency at which it revolves around the circle.

In our nuclear particle model, a photon rotating around the neutrino generates a magnetic field. As we saw in Chapter II, a photon captured by a singularity generates a rotating charge. A rotating charge creates a current and, therefore, a magnetic dipole. Nuclear particles exhibit magnetic fields that are inversely proportional to their mass. For example, the electron, with a mass that is about two thousand times smaller than that of a proton, has a magnetic field that is hundreds of times stronger than the proton field. The beauty of our model is that it can easily explain the reason for this peculiar particle behavior.

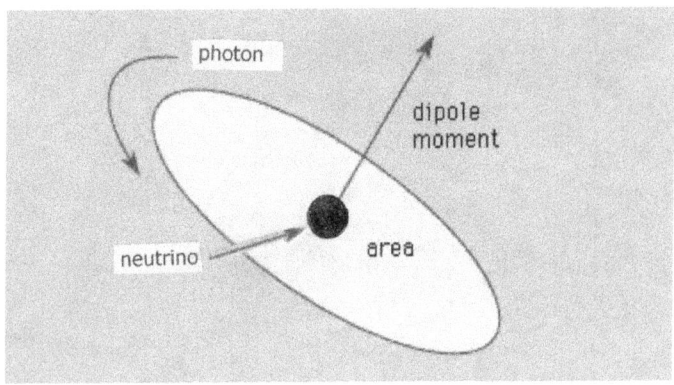

Figure 4.2- Electron Magnetic Dipole

At this point, we start analyzing the characteristic of the
electron.

Figure 4.2 depicts the dipole of an electron. In order to
compute the intensity of the magnetic field of the electron,
we have to find the area and the current. To find the area,
we assume that the photon depicted in *Figure 4.2* carries all
the energy of the electron. We will ignore the neutrino
mass in the center since it is known to be very small. We
know that the perimeter length has to be the photon
wavelength λ multiplied by an integer in order for the wave
to be a standing wave. We assume, first, that the integer is
1 and, therefore, the perimeter equals one wavelength λ.
Then we proceed with all the calculation as shown in
Appendix A and we find that the calculated figure for the
magnetic moment of the electron is consistent with the
present accepted values. The figure we have computed is
only one part in one thousand smaller than the actual
magnetic field of the electron. The difference can be
explained by the fact that in the equations we did not take
into consideration the mass of the neutrino; later on, using
more accurate calculations, our model will show the mass
of the neutrino to be less than one millionth the mass of a
proton. This mass is within the limits of the Oscillating
Neutrino Theory. The assumption we made, to use the
integer 1 for the number of wavelengths in our calculation,
has proved to be correct.

In our small black hole model, any time a photon (gamma
ray) is captured by a singularity, an anomalous
electromagnetic wave is generated; we call it anomalous
because the inner wave, being smaller than the outer wave,
creates a field imbalance. The net result of this imbalance is
that particles have the equivalent of a single charge rotating

at the speed of light around the singularity. This charge can be positive or negative, and its intensity is always the equivalent of the charge of the electron. (The explanation can be found in the Chapter X). The same effect happens in all the nuclear particles with the exception of the neutrino. The reason that the neutrino does not have a magnetic field is easily explained by our model. Our model shows the neutrino in the center of the particle. Whenever there is no photon associated with the neutrino, there is no rotating charge and, therefore, no magnetic field. No magnetic field associated with the neutrino has ever been found, even after repeated experiments with the most accurate and sensitive instrumentation available. This fact is one of the most important facts that our theory is based on.

At this point, we need to introduce a new nuclear particle that is not part of the stable particles family. This particle is the **muon**. Muons are created when cosmic rays collide with the Earth's upper atmosphere. Their life span is about 2.2 μsec. In the table in the following page, we compare certain characteristics of the muon and the electron.

If we divide the mass of the muon by the mass of the electron, we get a ratio of 206.8. If we divide the magnetic moment of the muon by the magnetic moment of the electron, we get a ratio of 1 / 206.8. From this result we can deduce that the magnetic moments in these two particles are inversely proportional to their masses. The greater the particle mass, the smaller the magnetic moment.
In Appendix A we compute the muon magnetic moment and the magnetic moment general formula.

$$M = q \, h \, / 4 \, \pi \, m$$

This general formula clearly indicates that the magnetic moment is inversely proportional to the mass of the

Particle	Mass(Mev)	Magnetic Moment(JT-ⁱ)
Electron	0.511	-928.48
Muon	105.66	-4.490

Magnetic Moment(JT^{-1}) 10 $^{-26}$

particle. Therefore, we have found the solution to our magnetic field enigma. When we apply our formula to the muon, we compute a magnetic moment that is only a few thousandths off the actual value. Our model has enabled us to solve the enigma that the smaller particles have a stronger magnetic field than the larger particles; furthermore, it has provided the means for computing the magnetic dipole associated with these particles with great accuracy.

In general, the electron and the muon are classified as part of the **lepton** family, while the proton and the neutron are classified as part of the **baryon** family. In the next chapter, we'll explain the differences between these two families.

Our model explains clearly how the magnetic moment of a nuclear particle is created and therefore we considered the **case** presented in this chapter one of our strongest cases.

[33]

Chapter V

The Proton

In 1918, the scientist Ernest Rutherford noticed that when helium nuclei were shot into nitrogen gas, his detectors showed the presence of hydrogen nuclei. Rutherford determined that the only place this hydrogen could have come from was the nitrogen, and, therefore, nitrogen must contain hydrogen nuclei. He suggested that the hydrogen nucleus, which was known to have an atomic number of 1, was an elementary particle. This particle was later called the proton. The proton and the neutron belong to the baryon family, while the electron and the muon belong to the lepton family. One of the major differences between these two families is the ratio of their mass to the magnetic moment.

NEUTRINO CASE No. 3

In this case we discuss what we call the "**proton problem**." This problem concerns the computation of the proton magnetic moment. The formula we derived in the previous chapter for computing the leptons magnetic moment does not work for the proton. In the previous chapter we derived the magnetic moment of the electron and the muon, and, in general, for the lepton family to which these two particles belong.

Unfortunately, when we apply this formula to the proton, the value we obtain for the proton magnetic moment is about three times smaller than the actual value. This is one of the reasons that the proton is classified as part of the baryon family. The following table shows the actual values. The mass of the proton is 1836 times the mass of the electron. The expected value of the proton moment, according to our formula, should be 0.505 instead of 1.410. Therefore, the actual proton magnetic moment is 2.79 times bigger than our expected value. In order to solve this problem we have to take into consideration the following observations:

Particle	Mass(Mev)	Magnetic Moment(JT-^{1})
Electron	0.511	-928.48
Proton	938	1.410

Magnetic Moment (JT^{-1}) 10^{-26}

- The charge of both the electron and proton is the same; therefore, the magnetic field difference is not due to the charge.

- Considering the electron and proton mass difference, the proton magnetic moment should have been .5 instead of 1.4.

- When we look at other particles in the electron family, such as muons, we find that muons are more penetrating than other particles of about the same mass that are

[35]

generated from the collisions of protons (**pions**). Muons, with approximately the same mass as pions, are capable of penetrating more when they are shot with the same speed of impact into a metal target. This would imply that muons have a smaller cross section.

- High-energy electron collisions give origin to unstable particles that are the same as the ones generated in proton collisions. This tells us that electrons and protons have structures that can be easily modified to resemble one another.

- Electrons have been created in laboratory experiments by colliding photons against photons. The same results are assumed to be possible for the creation of protons. This tells us that the basic building block of both particles is the photon.

From these observations, we take into account that both the electron and the proton, during collisions, yield the same components. At the same time both the electron and the proton require the same components when generated in a laboratory experiment. From this we can deduce that their structural difference must be minimal. This can be explained by a difference in their cross section. The cross section, not being equal, explains the difference of the magnetic moment and of the penetration capability. The solution to our problem is to find a factor equal to **2.79.** This means that we need to find a new particle configuration that is the same mass of the lepton and will provide an area that is 2.79 times bigger and, therefore, a magnetic moment that is 2.79 stronger. The first thing we try, since the proton cross section is about three times bigger than that of a lepton of the same mass, is to draw a picture with three wave cycles as in *Figure 5.1*.

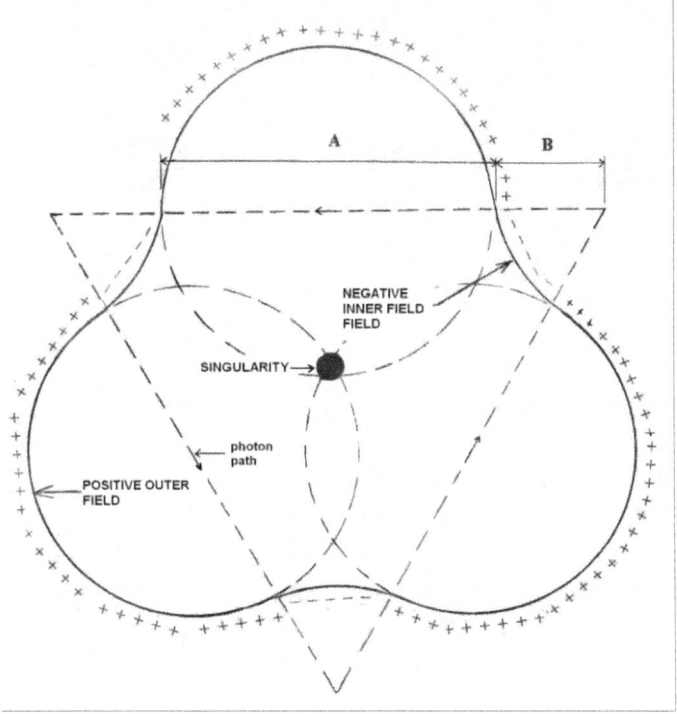

Figure 5.1 - The Proton Model with Polarized Photon Electrical Field

To draw this picture, first we draw an equilateral triangle, and then, in the center of each side of the triangle, we make a circle of diameter A. This is a three-wave configuration and according to the theory explained in the previous chapter, 120 degrees of each inner semi-wave of the polarized photon has to be absorbed by the singularity in the center. In *Figure 5.2-* when we cut off the black portions of the wave and bend the remaining wave in a circle, we obtain what is shown in Figure 5.1.

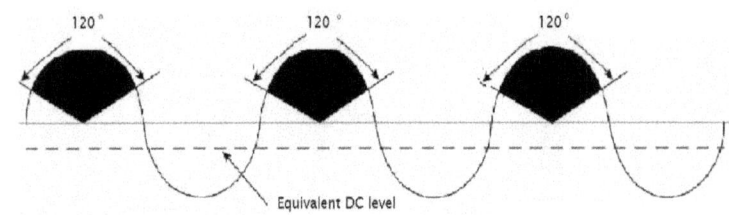

Figure 5.2- Parts of Standing Waves Captured

Assuming the top semi wave to be the negative field, when we subtract the dark portion, it leaves only 60 degrees for this semi-wave, as shown in *Figure 5.1*. Since between the 180 degrees of the outer semi-wave and the 60 degree of the inner semi-wave there is a ratio of 3 to 1, we select the segments A and B to have the same ratio. When we compute the area of the new proton model, we find it to be exactly 2.79 times larger than a single circle, representing the area of a lepton of the same mass.

One of the questions we ask is: Why do electrons and protons have different models? When we compare the two models, the proton diameter is about 1000 times smaller than the electron diameter. This means that the proton waves have to exist in much closer proximity to the singularity. In the three-wave model for the proton there appears to be more separation between the waves and the

singularity in the center. This model appears to be the only solution for a particle of the proton mass that is stable.
 The models that we have created appear to be very sensitive to the wavelength of the photons. Only certain wavelengths generate stable particles. Other wave lengths create particles that decay in a very short time because they are not compatible with the gravitational field of the neutrino. Another cause for particle decay can be that the beginning and the end of the waves are not in phase with each other, preventing the wave from becoming a permanent **standing wave.**

In each model (electron and proton), the singularity absorbs a portion of the photon electromagnetic field. The absorbed portions represent the particle potential energy. In the case of the electron, half of the wave is absorbed. In the case of the proton, only 120 of the 360 degrees are absorbed, or one-third of each of the three waves. With the potential energy equal to one-half to one-third of the total energy of the particle, these models are very stable. These models compare very favorably with respect to the nuclei of the most stable elements, where the potential energy is only a small percentage of the total energy.

 In this chapter we have described the particles as two-dimensional models; in reality, due to a continuous spinning, their shape is spherical (three-dimensional) instead of circular. Their shape gets flatter and flatter when they are exposed to an accelerating electric field.
We surmise that the only two stable charged particle in the Universe, the electron and the proton must have a considerable difference in their masses, otherwise they

would self destruct whenever they would get near each other like matter and antimatter.

We do not address the neutral baryons and leptons in this chapter because their models are more complex, since they probably include two waves of opposite polarization.

Chapter VI

The Friedman, Taylor and Kendall Experiment

NEUTRINO CASE No. 4

When the quarks theory was originated in the year 1963, there could have been some suspicion on its validity, since it was started as a method to catalog the more than 100 stable and unstable particles already known. At that time, Murray Gell-Mann, the scientist credited for the theory, called the quark a mathematical entity instead of a real entity, probably thinking that the existence of quarks would never be proven. Around 1970, Dick Taylor, Henry Kendall and Jerry Friedman using high energy electrons were able to look inside the structure of the protons and neutrons. Since they were unable to find any solid masses, they concluded that any component inside these particles had to be very small and similar to the point like ½-spin structure of the electron and neutrino. The findings that got the most attention were generated when they saw fractions of the electron charge in their experiments. These findings created much enthusiasm in the scientific community because they were assumed to be the proof of the validity of the quarks theory which assigns to quarks 1/3 or 2/3 electron electrical charges.

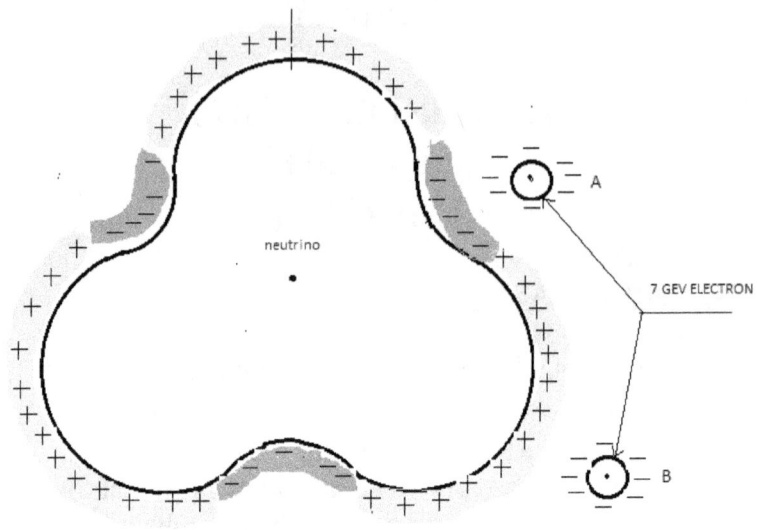

**Picture 6.1 - Scattering of Proton and 7.0 Gev Electrons-
Drawn to Scale**

But the quarks structure formulated for the protons and
neutrons is not the only possible structure that would match
the results recorded by the Taylor, Kendall and Friedman
team. As matter of fact we are confident that the new
particle model we introduced about six years ago would
satisfy the two major findings of the Taylor, Kendall and
Friedman experiment. First, it would satisfy the point like
½-spin requirement because it has a point like ½-spin
neutrino in the center of the structure. Second, this new
particle model would show fractions of electron charges
when analyzed using high energy electron beams.

In Picture 6.1 we simulate the Taylor, Kendall and

Friedman's experiment. We show a proton being
bombarded by 7 Gev electrons. Both the proton and the
electrons are depicted according to our particle theory.
The7 Gev electrons are shown to be approximately 7 times
smaller than the 938Mev proton. Normally electrons at rest
are about 1000 times the size of a proton, but in this picture
they are depicted smaller since these are high energy
electrons with seven times the proton rest mass. The
proton, when observed from a distance, shows a charge
equivalent to 1.0 electron charge but when observed in
proximity it shows a totally different picture. The charge of
the proton is formed by three cycles of an electromagnetic
wave warped by the effect of the singularity in the center.
There are three 180 degrees of positive wave and three 60
degrees of negative wave in this model. Part of the negative
waves has been absorbed by the neutrino. From the picture,
subtracting the positive waves from the negative waves, we
obtain a total of 360 degree of positive wave and the final
effect is the 1.0 electron positive charge. We can see from
the picture that the electron in position A will be scattered
by the negative field of the proton and that the electron in
position B will be scattered by the positive field of the
proton. The scattering force on electron B is probably twice
as strong as the force on electron A. This could have led the
scientist to believe that there were two types of quarks, one
with 2/3 positive electron charge and one with 1/3 negative
electron charge. From this experiment we cannot say that
our model is a better model when compared to the Quarks
model. The only thing we can say is that, the belief in the
existence of quarks based on the result of this experiment,
could be misleading.

CHAPTER VII

THE PROTON SPIN CRISIS

NEUTRINO CASE No. 5

One of the properties of nuclear particles that can be measured is their **spin**. In our theory, we can explain quite simply, with a few sentences, why so many of these particles have the same spin. In the center of all nuclear particles there is a neutrino. The neutrino is a ½ spin particle. All the stable particles, since they are made by photons captured by neutrinos, have ½ spin. The gamma rays or photons circling the neutrino are polarized **0** spin photons and therefore have no impact on the total amount of the particle spin. We can say that, when we observe an electron or a proton spin, we are really observing the spin of the neutrino in its center. The simplicity of this theory makes even more sense when we consider that the proton mass is about two thousand times greater than the electron mass and this difference does not have any impact on their spin. In our model the mass is carried by the photon and the spin stays the same no matter the mass of the particle. Unfortunately things are much more complex in the present particles Standard Model. Though both electrons and protons have exactly the same spin, their nature is portrayed completely differently, since the proton structure includes three quarks. Each quark has its own spin and the sum of their spin equals the total particle spin. This explanation was considered valid for about 25 years until 1988.

In 1988, more than 20 years ago, the nuclear physicists found themselves in the middle of a crisis. This crisis was due to the results of a proton- muon scattering experiment that showed that little or no spin of the proton could be attributed to the quarks inside it. Since previously the quarks were considered the generators of the proton total spin, this was a major upset for the Standard Model theory. Soon after, many papers were written explaining these results with different theories. The accepted theory was that the gluons, another component of the proton according to the Standard Model, were the carriers of the missing spin. This solution, with six different components of the proton sharing its ½ spin, is, in our opinion, too complicated and totally lacking in elegance. Elegance is the most recurrent theme that we find in the Universe and this solution had none. Even the fact that many papers were so readily written to explain away the problem is not surprising considering the thousands of scientists that were working on it and the billions of dollars that were being spent on projects based on the quarks model. Nobody wanted this bonanza to be stopped because, suddenly, this theory had achieved a dubious credibility in view of the new findings. After all, in the history of mankind, nobody ever wanted to talk or write himself out of a good job.

To explain our last sentence we need to make an analysis of some historical facts. In the five decades following 1930, the nuclear particles science achieved its peak. After the discovery that the uranium atom could be split, numerous scientists throughout the world started working on the possible implication of such finding. In 1941, Enrico Fermi achieved continuously sustainable uranium fission building a nuclear reactor at the University of Chicago. This success made it possible for an even larger team of scientists, in the

Manhattan project, to deliver the first atomic bomb in 1945. After the atomic bomb, the 100 times more powerful hydrogen bomb was developed. On the civilian front, nuclear reactors began generating electricity that promised unlimited, clean and cheap energy.

In this environment, considering all the military and civilian applications of nuclear science, large budgets were appropriated for the research in this field. The cold war helped to expand this field of research because after the atomic and hydrogen bomb, nobody knew what other terrifying weapon could be developed. The cold war ended in 1989 with the fall of the Berlin wall and its fall marked the end of the paranoia about new nuclear weapons. In 1994 the largest particle accelerator project located in Texas was cancelled, and one of the factors was probably the end of the cold war.

Going back to the proton crisis, we can see that in 1988 there was a lot of money being poured into the nuclear particles field and nobody wanted to stop this bonanza. If no big money had been involved probably this crisis would have given enough incentives to make some changes to the Standard Model, but unfortunately this was not the case.

Chapter VIII

The Neutron Model

NEUTRINO CASE No. 6

The neutron has about the same mass of the proton but carries no charge. According to our theory, photons, when captured by a singularity, can have a positive or negative charge. In order for a particle to have no charge it has to have a minimum of two photons rotating around a singularity. One that creates a positive charge and one that creates a negative charge. The two photons create equal charges of opposite polarity and the resulting particle is a neutral particle. The charges are always equal whenever any photon is captured by a singularity no matter its frequency (see Chapter X). Charges have to be of the opposite sign to produce neutrality. This is not a problem because the polarity depends only on whether the positive or negative semi-wave is absorbed. In *Figure 8.1* we show the singularity surrounded by two photons, one with a negative charge and one with a positive charge. Any particle with this structure is a neutral particle. The particle shown in *Figure 8.1* belongs to the lepton family because the photons that circulate around the singularity have only one wave length.

The neutron belongs to the baryon family; therefore, the model in *Figure 8.2* is more appropriate for a neutron since in the baryon family the captured electromagnetic wave has three wave lengths. This concept was shown before in the proton model.

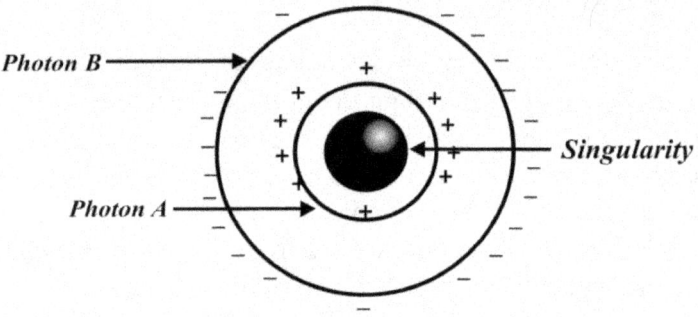

Figure 8.1 - Structure of a Neutral Particle.

The neutron always decays into a proton and an electron. If we build the neutron model using two photons, one with the energy of an electron and the other with the energy of a proton, then the resulting magnetic moment is totally erroneous. Since we know the neutron mass and magnetic moment, by solving a system of two equations, we can compute the mass of the photons that constitute our neutron model.

[48]

We can call these photons, particles because they are captured photons and therefore have all the attributes of particles. In reality they are photons whose energies can be translated directly into masses. In the following table we show that we cannot match the mass and magnetic moment of the neutron when we add a proton and an electron.

Particle	Mass(Mev)	Magnetic Moment(JT⁻¹)
Neutron	939.57	-0.966
Electron	0.511	-928.48
Proton	938	1.410
Electron + Proton	938.78	-927.07

Magnetic Moment (JT^{-1}) 10^{-26}

By adding the masses and the magnetic moment of particles A and B, we can match the mass and magnetic moment of the neutron. This same result cannot be achieved adding the electron and proton values shown in the table above.

In the following table we show the two particles (or the two photons) that match the neutron total mass and the neutron magnetic moment; we have called them particle A

[49]

Particle	Mass(Mev)	Magnetic Moment(JT-¹)
Neutron	939.57	-0.966
Particle A	574.6	1.681
Particle B	365	-2.646
Particle A+ Particle B	939.6	-0.965

Magnetic Moment (JT^{-1}) 10^{-26}

(or photon A) and particle B (or photon B).

Therefore, we find that one of the solutions to the structure of the neutron is the presence of two photons of opposite polarity, one with the energy of about one third and the other with the energy of two thirds of the neutron total mass. **Note that in *Figure 8.2* the outside electromagnetic wave presents a negative charge to the outside world. With this type of configuration the neutron should have an attraction force for protons located at a range equal to its diameter. At farther distances, this electrical field would be neutralized by the positive charge of the inner electromagnetic wave. This type of configuration would help the protons and neutrons to bind together in the nucleus of an atom and would explain why the so called "strong forces" have such short range.**

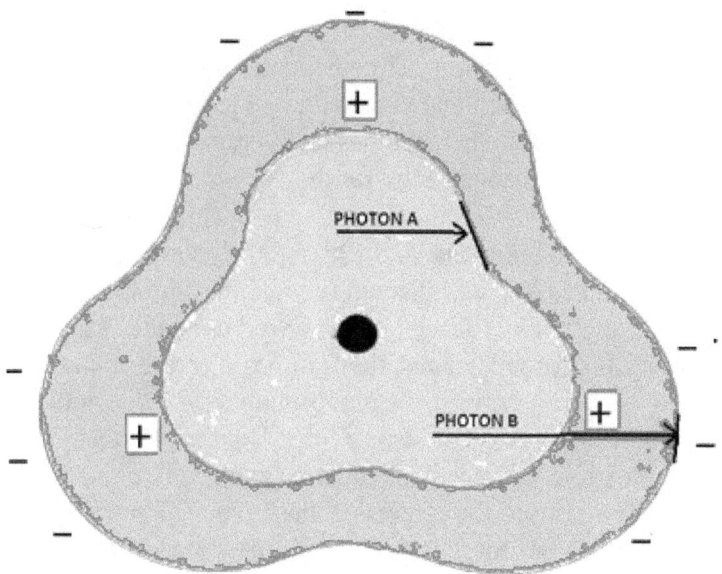

Figure 8.2 - The Neutron Model with Two
Overlapping and Opposite Polarized Fields

The neutron is probably a more complex particle than the
one shown in our model.

A continuous exchange of energy between particle A and
particle B is very likely. In the nucleus of an atom, these
exchanges of energies would also include the protons,
generating the cohesion force of the nucleus.

We can surmise that the neutron is not a stable particle
because, when it is exposed to a gravity field or an ocean of
free neutrinos (as described in the following chapter), one
of the surrounding photons may be captured by a free
moving neutrino causing the neutron to decay.

neutron + neutrino → proton + electron

In this particles collision, a free neutrino from the ocean of neutrinos collides with a neutron and generates a proton and an electron. The neutron has an average life of 760 seconds on the surface of the Earth.

At this point, since we have discussed the proton and the neutron models, we will discuss the nucleus and the so-called strong nuclear force (also referred to as the "**strong force**"). This force is one of the four basic forces in nature (the others being gravity, the electromagnetic force, and the weak nuclear force). The strong force has the shortest range, meaning that particles must be extremely close before they react to the effects of the force. The main job of the strong nuclear force is to hold together the subatomic particles of the nucleus. Protons carry a positive charge, and neutrons carry no charge. These particles are collectively called nucleons. If we consider that the nuclei of all atoms except hydrogen contain more than one proton, and each proton carries a positive charge, why do the nuclei of these atoms stay together? The protons must feel a repulsive force from other neighboring protons. This is where the strong nuclear force comes in. The nucleons, though, must be extremely close together in order for this exchange to happen. The distance required is equal to about the diameter of a proton or a neutron. If a proton or neutron gets closer than this distance to another nucleon, the exchange of photons occurs, and the particles stick to each other. If they can't get that close, the strong force is too weak to make them stick together, and other competing forces (usually the electromagnetic force) influence the particles to move apart. A particle must be able to cross this barrier in order for the strong force to "glue" the particles together.

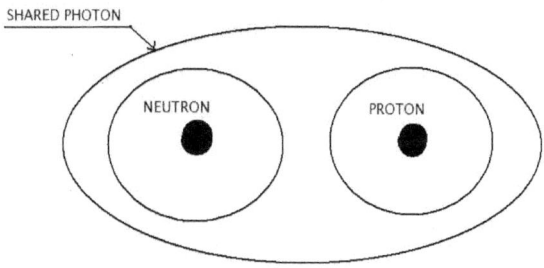

Figure 8.3 - Nucleons Exchange of Photons

In the case of approaching protons, the closer they get, the more they feel the repulsion from the other proton/nucleus (the electromagnetic force). As a result, in order to get two protons/nuclei close enough to begin exchanging photons, they must be moving extremely fast (which means the temperature must be really high), and/or they must be under immense pressure so that they are forced to get close enough to allow the exchange of photons to create the strong force (*Figure 8.3*).

One thing that helps reduce the repulsion between protons within a nucleus is the presence of any neutrons. Since they have no charge, they don't add to the repulsion already present, and they help separate the protons from each other so they don't feel as strong a repulsive force from any other nearby protons. Also, the neutrons are a source of more strong force for the nucleus since they participate in the photons exchange. These factors, coupled

[53]

with the tight packing of protons in the nucleus so that they can exchange photons, create enough strong force to overcome their mutual repulsion and force the nucleons to stay bound together. This is why it is easier to bombard a nucleus with neutrons than with protons.

In this chapter and in the following chapters, we cover all the four fundamental forces of physics. All these forces, according to our theory, are just different ways that neutrinos and photons interact with each other. The photon is considered to have a spin equal to 1, which is in agreement with other nuclear physics theories that require any particle related to the exchanges of these forces to have a spin equal to one. A polarized photon has a spin equal to 0 but, it still retains all the properties of the non polarized photon. In the Standard Model of nuclear physics these pseudo particles are called Bosons. They are called pseudo particles because of their extremely short life span. Their life span is so short that their existence can only be surmised by the observation of laboratory data. In our theory we replace bosons with high energy photons.

Chapter IX

The Supernova Enigma

NEUTRINO CASE No. 7

In the recent past, an astronomical event took place that we consider important in the development of our theory. In the year 1987, in the galaxy known as the Large Magellanic Cloud, located some 170,000 light years away from Earth, a star exploded into a supernova. The explosion converted most of the mass of the original star into energy.
During supernova events such as this, protons, neutrons, and electrons are converted into energy. This mass energy conversion creates byproducts: a wide spectrum of electromagnetic waves—from gamma rays to visible light rays and neutrinos. Most of the gamma rays never reach Earth because they interact with intergalactic matter. Neutrinos, however, are capable of traveling intergalactic space. In the case of the 1987 supernova, a shower of neutrinos was detected on Earth at about the same time scientists were able to see the visible light spectrum.

Due to the lack of accuracy in the instrumentation a maximum separation of a few hours could have occurred between the arrival time on Earth of the light spectrum and of the neutrinos. This means that the difference in velocity between the neutrinos and the speed of light, taking into account the distance of the event, could have been, at most, one billionth, but most probably the neutrinos were indeed traveling at the speed of light. Present-day physics does not allow any particle to travel at the speed of light, and the possibility of an arrival time delay has been used to explain that the neutrinos were not traveling at the speed of light. At this point, our ideas diverge from those of present-day physics. We believe that, in this case, the neutrinos were very possibly traveling at the speed of light.

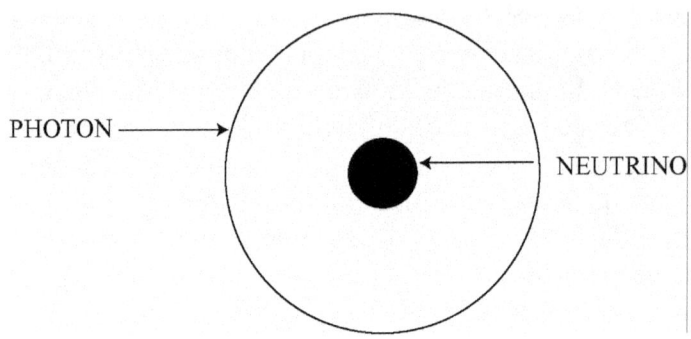

Figure 9.1- Black Hole Model of the Electron

Our small black hole model predicts that, in the case of transformation of particles into energy (as with a supernova) there will be a shower of free neutrinos and photons because these are the only components available in the stable particles, like the proton, the neutron and the electron, before their destruction. At this point we need to explain how it is possible for the neutrino to move at the speed of light when it has been proved that all the rest of the particles do not have this capability.

In order to understand this concept, we need to understand what happens to particles, such as electrons or protons, when they are accelerated using an electric field. Let us assume that the particle depicted in *Figure 9.1* is an electron being accelerated by an electrical field perpendicular to the page of this book. The electron will start moving in the same direction of the electrical field (toward the reader's face). The electron motion away from the page toward the reader will change the path of its photon. When the photon is part of a stationary electron, it moves in a circle around the neutrino. When the photon is part of a moving electron, it is unable to maintain the circular motion and its trajectory becomes a spiral. A viewer located on the surface of *Figure 91* will see the path of the photon as it is shown in *Figure 9.2.*

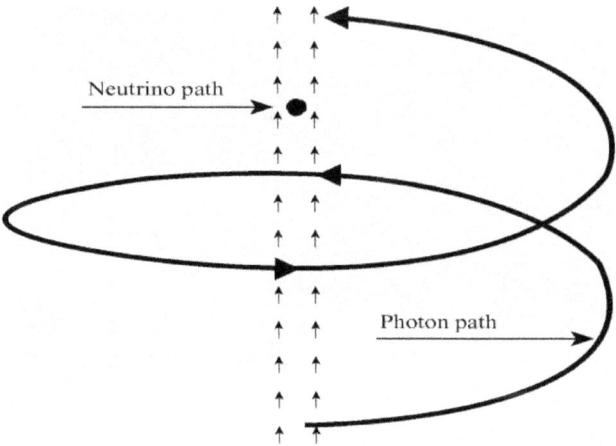

Neutrino path

Photon path

Figure 9.2 - Electron Black Hole in Motion

The two dotted lines in *Figure 9.2* depict the motion of the neutrino in a straight line in the middle of the spiral. We need to keep in mind that the neutrino and the photon are the only components of the electron. In this figure, the electron moves toward the top of the page, and the photon, while rotating around the neutrino, is moving in the same direction because it is an integral part of the structure of the electron. Therefore, the singularity or neutrino moves in a straight line while the photon moves in a spiral path. This figure explains why the electron cannot achieve the speed of light while moving toward the top of the page. Its main component, the photon, though moving at the speed of light, is forced into a time-consuming spiral path.

Figure 9.3- An Analogy

Figure 9.3 presents an analogy to help clarify this concept. A car is moving along a curvy road from location A to location B. Assuming that, starting from point A, the car is in a race competition with a light beam, and, assuming that the car itself is moving at the speed of light, we have no problem acknowledging that the car will lose the race. The light beam will win the competition because it can move in a straight line, while the car, like our photon in *Figure 9.2*, must move along a winding path.

Going back to our 1987 supernova, the neutrino shower that was observed on Earth was made up of free neutrinos—free neutrinos have no photons in orbits around them. These neutrinos should have been capable of traveling at the speed of light because they were not limited by spiraling photons. All the other particles, such as electrons and protons, are subjected to this limitation and are not capable of achieving the speed of light. In the next chapters we show that the neutrinos are just a different form in which the electromagnetic waves can manifest themselves and therefore are capable of travel at the speed of light

The model of the small black hole has allowed us to understand why particles in general cannot be accelerated to the speed of light and why the neutrino is the only particle that possibly can escape this constraint. This fact is not necessarily a violation of the Einstein's Theory of Relativity since the next few chapters will show the neutrino to be just another form of the electromagnetic wave.

Chapter X
The Charged Particle

<u>NEUTRINO CASE No. 8</u>

In this Chapter we investigate why some nuclear particles display an electrical charge and why this charge is always of the same intensity. To accomplish this task we'll use the model of the electron that we have already developed. In laboratory experiments, electrons have been synthesized by colliding photons against photons. The implication of these experiments is that there should be some similarities in the structure of the photon and the electron. Using this concept, we have developed a theory that enables us to find the size of the charge of the electron. Before we get involved with the task of computing the electron charge, we need to clarify our understanding of the photon.

The Photon

Many physicists believe that the photon has no dimensions. We diverge from this thinking because the photon is an electromagnetic field and an electromagnetic field, in order to carry energy, has to have a certain dimension. The most common photons are the ones in the visible spectrum, but the ones we'll discuss in this chapter are photons with much higher energy—the gamma ray photons. The only difference between these different types of photons is the frequency of their electromagnetic fields. The electrical E and magnetic H fields of the photons can be represented mathematically with the following equations:

$$E = E_{Max} \sin \{2\pi \upsilon \, [t - z \, (\mu \, \varepsilon)^{1/2}] \}$$

$$H = H_{Max} \sin \{2\pi \upsilon \, [t - z \, (\mu \, \varepsilon)^{1/2}] \}$$

The frequency of the wave υ is the only difference between the visible spectrum photons and the powerful gamma rays. A polarized photon is represented in *Figure 10.1* with the electrical field **Ex** along the x axis and the magnetic field **Hy** along the y axis, orthogonal to each other, varying sinusoidally along the time axis **t**. *Figure 10.1* shows only one wavelength cycle; the number of cycles that are in a photon is presently unknown.

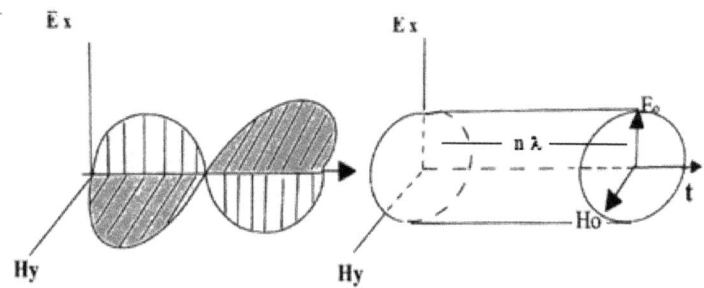

Figure 10.1
Electromagnetic Field

Figure 10.2
Equivalent DC Field

Energy is associated with every photon. To compute this energy, we need to find the root mean square value of **Ex** and **Hy**—the so called RMS values.

In *Figure 10.2* we redraw the photon in *Figure 10.1* by using **Eo** and **Ho**; these are the RMS values of **Ex** and **Hy**. **Eo** and **Ho** are shown in *Figure 10.2* as constants, and the electromagnetic field now looks like a solid cylinder. It is possible to compute the energy associated with an electromagnetic field by knowing the volume V occupied by the field.

At this point, we have to make some assumptions in order to compute the volume associated with the photon. The cylinder we observe in *Figure 10.2* has to have a length equal to a multiple of the wavelength λ or **nλ , where n is the number of cycles**. The base has a radius equal to $\lambda / 2$ since **Ex** and **Hy** stay polarized with a positive or negative field for the entire half wave cycle. The **Ex** and **Hy** travel at the speed of light in the X and Y directions, covering the distance $\lambda / 2$. Therefore the volume V can be computed:

$$\text{The volume V} = \text{length x base} = n\lambda\, \pi\, (\lambda / 2)^2$$
$$= n\pi\, \lambda^3 / 4$$

The total energy of the photon is stored in the volume of its electric and magnetic fields:

$$\text{PHOTON ENERGY} = (1/ 2\; \epsilon\, E_0^2 + 1/2\mu\; H_0^2)\; V$$

The amount of energy in the electrical and magnetic field is the same; therefore:

$$\text{PHOTON ENERGY} = \epsilon\, E_0^2\; V$$

Substituting for the volume V from the previous equality we obtain:

PHOTON ENERGY = (ϵ E_0^2) V = (ϵ E_0^2)n π λ^3 / 4

It is also known that photon energy = $h\upsilon$; therefore:

PHOTON ENERGY = (ϵ E_0^2) V = (ϵ E_0^2)n π λ^3 / 4

= $h\upsilon$

In the next paragraph we'll make use of this equality that we have derived for the photon energy.

The Electron

In Chapter II, we showed how gravity can bend the travel of the photon by acting on the photon's internal clock.

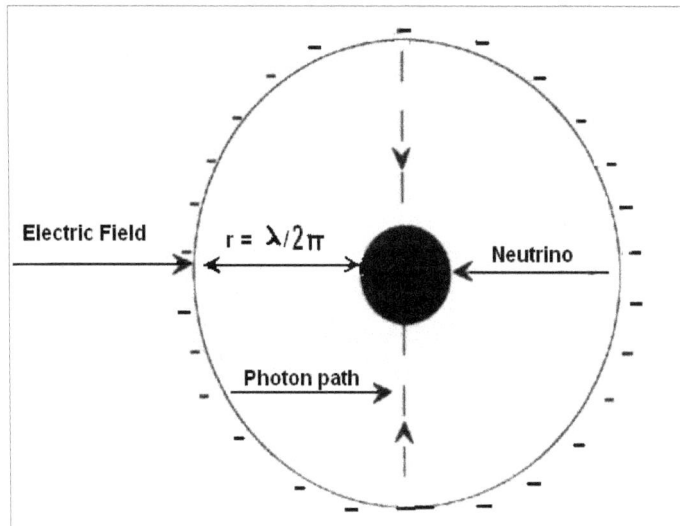

Figure 10.3- The Electron Model

In the case of the electron, the internal clock of the semi-wave that is closer to the singularity must be affected to the point that, for each rotation of the polarized photon around the neutrino, the two positive semi-waves are totally deleted.

The two negative semi-waves that have escaped the gravity field form a single standing wave circle as shown in *Figure 10.3.* The particle shown is made totally of the negative remaining fields and it is therefore negatively charged. As we already discussed in Chapter IV, the circumference of the electron is equal to one wavelength λ, and the radius is **$r = \lambda/2\pi$.**

The electrical field of the photon in *Figure 10.4*, after the photon is captured by the neutrino, has 180 degrees missing from each cycle of the inner wave (shaded area).

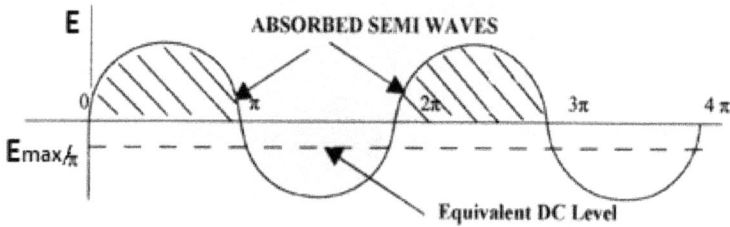

Figure 10.4 - The Electron DC Level

The resultant wave has a unique characteristic because now it has a negative DC component. This DC component creates what we call the electron charge. This DC level is computed in *Figure 10.4* integrating over the two cycles of the wave or a period 0 to 4π. Excluding the shaded areas, the DC level equals to E_{Max} / π and equals the E_0 of the previous computations. Using the models for the electron and the photon that we have developed, in the next paragraph we will compute the charge of the electron.

The Experiment

If we perform an experiment in which a beam of photons collides with another beam of photon, no electrons are generated until the energy of the colliding photons reaches .511 MeV. At this energy level, electrons can be detected. This energy is the equivalent energy of the rest mass of the electron. The transformation of the photon into a particle is coherent with our model. During this transformation, there was a change in the photon traveling pattern: Before the collision, the gamma ray traveled in a straight line. After the collision, it traveled in a circular path. We hypothesize that neutrinos located in the proximity of the colliding gamma rays have captured them, putting them into orbits and therefore causing this transformation. This theory assumes that large quantities of free neutrinos are present everywhere in the Universe. In Chapter II, we showed how gravity could bend the travel of the photon by acting on the photon's internal clock. In the case of the electron, the internal clock of the semi-wave that is closer to the singularity must be affected to the point that, for each

rotation of the photon around the neutrino, the positive
semi-waves are totally deleted.

The two negative semi-waves that have escaped the gravity
field form a single standing wave as shown in *Figure 10.3*.
The particle shown is made totally of the negative
remaining fields and it is therefore negatively charged. As
we already discussed in Chapter II, the circumference of
the electron is equal to one wavelength λ and the radius is r
$= \lambda/2\pi$.

The Electron Charge

To compute the charge of the electron, first we need to
compute the volume of the photon electric and magnetic
fields before they were captured by the neutrino. We know
from *Figure 10.2* that the length of the field is **n**
wavelengths or **n** λ . The radius of the cross section is
equal to $\lambda / 2$. This dimension appears to be reasonable
because during the duration of one half wave, a photon that
travels at the speed of light in the **t axis** direction, will have
its electric field and magnetic field travel the same distance
in the **X** and **Y axis** directions. Since the number of wave
cycles that make up a photon is unknown, we assume the
value for **n**, the number of cycles, to be the integer **4**. This
means that the captured photon that generates the electron
is originally composed with four wavelengths [1]. Using the
RMS values of the electromagnetic field, we can represent
the photon by a solid cylinder of **length 4λ** and **radius λ/2**.
Then we compute the value of the volume V of the photon
using the formula for the volume of a cylinder:

$$\mathbf{V} \; = \; \textbf{length} \; \; \mathbf{x} \; \; \textbf{base} \; = \; \mathbf{4\lambda \, \pi \; (\, \lambda \, / \, 2 \,)^2 = \; \pi \, \lambda^3}$$

The total energy of the photon E $_{field}$ is stored in the volume of its electric and magnetic field :

$$E_{field} = (\,1/2\ \epsilon\ E_0^2 + 1/2\mu\ H_0^2\,)\ V = \epsilon\ E_0^2\ V$$

where E_0 and H_0 stand for the intensities of the electric and magnetic fields and ϵ and μ for the dielectric constant and permeability of vacuum.

Substituting for the computed value of the volume we obtain:

$$E_{field} = (\,\epsilon\ E_0^2\,)\ V = (\,\epsilon\ E_0^2\,)\ \pi\ \lambda^3$$

(Equation10.1)

The total energy stored in the electron electromagnetic field, plus the potential energy of its gravity field, must equal the energy of the photon before the collision that produces the electron. This energy equals the Planck constant multiplied by the frequency of the wave.

$E_{field} + E_{potential} = h\ v$; Since the positive semi-waves have been absorbed, or 50 percent of the original magnetic field has disappeared, we can say that the potential energy equals the field energy, and therefore, going back to *Equation 10.1* we get:

$$E_{potential} = E_{field} = \epsilon\ E_0^2\ \pi\ \lambda^3 = h\ v\,/2 =$$
$$= h\ c\,/\,2\ \lambda$$

[67]

The intensity of the electric field is:

$$E_0^2 = (h\,c\,/\,2\,\lambda)\,/\,\epsilon\,\pi\,\lambda^3$$

$$E_0 = (h\,c\,/\,2\epsilon\,\pi)^{1/2}\,/\,\lambda^2 \qquad \textit{(Equation 10. 2)}$$

This computed intensity of the electrical field is the intensity we find on the surface of a sphere of radius $\lambda/2\pi$, which is the effective radius of the electron. (*Figure 10.5*) This was the same radius used to compute the magnetic moment.

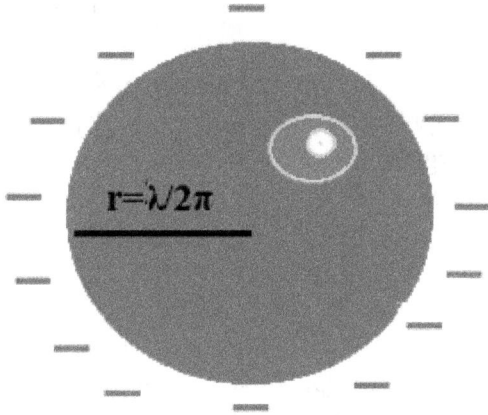

Figure 10.5 - The Electron Charge on the Surface of the r = $\lambda/2\pi$ Sphere

The two outer semi-waves combined form the equivalent of a sphere due to a rotation around their axis. This rotation gives our electron model a third dimension. Since the sphere has, on the surface, an electric field of intensity Eo, using the Gaussian theorem we can compute the enclosed charge to be:

$$q = 4\pi \ r^2 \ \varepsilon \ E_0$$

Then we substitute the effective radius for r

$$r = \lambda \ / \ 2\pi$$

And we obtain:

$$q = 4\pi \ \varepsilon \ (\lambda \ /2\pi)^2 \ E_0$$

Where $4 \ \pi \ r^2$ is the surface of the sphere and ε is dielectric constant of vacuum.

$$q = \varepsilon \ (\lambda)^2 \ E_0 \ / \ \pi$$

substituting for E_0 (Equation 12.2) we obtain:

$$q = (h \ c \ \varepsilon \ / \ 2\pi^3)^{1/2}$$

$$q = 1.68 \ 10^{-19} \ \textbf{coulombs}$$

This charge is only 5 percent different from the actual electron charge of $1.60 \ 10^{-19}$ coulombs and this supports the validity of our primitive model.

With a computer simulation model a more accurate result could be possible.

It is important to note that the final formula for the charge is independent of the wavelength; therefore, no matter the particle mass the charge is always the same and this is consistent with the observed nature of the particles.

$$q = (h \ c \ \varepsilon \ /2 \ \pi^3)^{1/2}$$

This formula is extremely important because the charge is independent of the wavelength. All polarized photons rotating around a singularity create the electrical charge of an electron. This explains why no particle with a fraction of the electron charge has ever been observed.

The Proton Charge

To show that the proton charge is the same as the electron we review *Figure 10.6.* In the picture of the proton, we have the positive outer field that covers three cycles of 180 degrees for a total of 540 degrees.

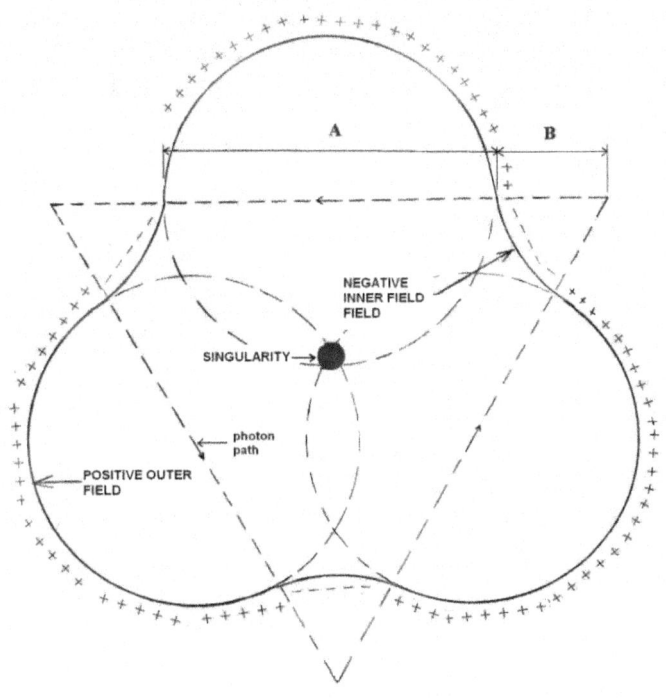

Figure 10.6 - The proton charge

The negative inner field has three cycles of 60 degrees for a total 180 degrees. When we subtract the negative field from the positive field, we obtain a total of 360 degrees of positive field. This is the same result as Figure10.3 for the electron, only difference is that the charge is of the opposite sign. To be notice is the fact that the actual photon path is an equilateral triangle for the proton and a straight line for

[71]

the electron. The photon of the electron moves back and forth along the same line.

If we consider a hypothetical situation in which a lepton (electron) has the same mass of a baryon (proton), in this case the photon needs to complete two cycles (lepton model) in order to cover the same path length of the baryon model.

     1. The four cycle wave appears to be confirmed by the fact, that leptons and baryons of the same masses, to comply with the computations of Chapter V, need both to be composed by a four cycles wave.

Chapter XI
The Dark Matter and the Gravitational Theory

NEUTRINO CASE No. 9

In the previous chapters we have discussed how neutrinos can generate different types of particles by capturing photons. In this chapter we'll look into how large bodies in the Universe are affected by the gravitational force and how this force can be generated by neutrinos even at great distances by bending the path of photons through the warping of time and space. From a distance, neutrinos can affect photons freely moving through space like the Sun rays, or they can affect photons that are integrated into particles belonging to a celestial body such as the Moon. The exchange of gravitational forces is always and solely due to the interactions between the neutrino and the photon. Our theory now moves from the microscopic to the macroscopic observation.

According to our theory, the Universe is totally permeated by an ocean of small singularities or *"free- neutrinos"* (neutrinos without photons). This ocean can be called the *"Dark Matter"*. Recently, astrophysicists have agreed that in the Universe a large percentage of all the matter is invisible matter or Dark Matter. The neutrinos or singularity, being small black holes, fit very well in the invisible matter or Dark Matter picture. The density of this ocean of singularities varies in proportion to the amount of *"visible matter"* present in any given corner of the Universe. The lowest density of this ocean is in intergalactic space. The highest density of this ocean is in the center of every galaxy—where a gigantic black hole

[73]

resides. In the next paragraphs, we are going to refer to these singularities or neutrinos as *free- neutrinos* to differentiate them from the neutrinos that are at the center of particles as in the electron, the proton and the neutron. Gravitational fields affect all matter with the exception of the *free- neutrinos* because, being singularities without a photon, they cannot affect each other clock, and therefore cannot cluster together on their own. The neutrinos interact with photons or other nuclear particles acting on their time clock. [1] Such interaction would be impossible among *free- neutrinos* because they are singularities and lack an internal time clock.

As we discussed previously, particles are photons captured by neutrinos, and objects are made by the amalgamation of certain quantities of these particles. The quantity is proportional to the mass of the object. Therefore the Universe is made of a combination of *free- neutrinos* and neutrinos surrounded by photons. The *free- neutrinos* represents the dark matter, and the neutrinos with photons represent the visible matter present in the Universe.

At this point we will describe the interaction among the *free- neutrinos* and the visible matter, which is made up by the electrons, the protons and the neutrons. In the following table we use the abbreviation PEN, combining the proton, electron, and neutron in the same category.

PEN	↔	PEN	Reciprocal attraction
PEN	→	*free- neutrinos*	One way attraction
free- neutrinos	−	*free- neutrinos*	No attraction
PHOTON	→	*free- neutrinos*	One way attraction

All the gravitational attractions in the above table are generated by neutrinos. The neutrinos located in the middle of the PENs and also the *free- neutrinos* both, generate a gravitational attraction by the warping of the **time** and therefore cause the bending of the photon paths.

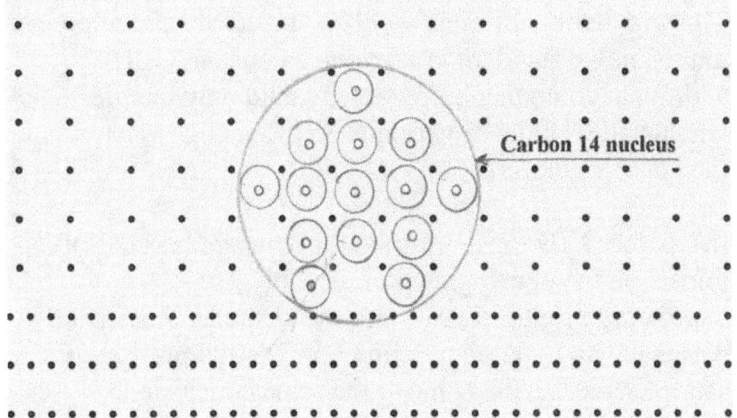

Figure 11.1- A Carbon Nucleus in a free- neutrino Field

The gravitational force probably has **one more component** that could manifests itself in very high density *free-neutrino* fields. This other component can be explained in the following way. *Figure 11.1* depicts a carbon nucleus that includes 14 nucleons, each with photons rotating around a neutrino (smallest circles). [2] The carbon nucleus is inside a *free- neutrinos* field (black dots) that increases in density toward the bottom of the picture. The *free neutrinos* continuously move in and out of the nucleus. When a *free-neutrino* encounters a PEN (neutrinos with photons), it puts the PEN in an ambiguous situation that could result in the

PEN exchanging the neutrino in its center. Because of the probabilistic effect of the transfer of these photons among neutrinos, the photons, and therefore the nucleus, will travel toward higher-density–*free*- *neutrino* fields. In *Figure 11.1*, the carbon nucleus is attracted toward the bottom of the picture. As the carbon nucleus moves toward the higher-density neutrino field, the exchange rate increases and generates an increasing gravitational force. This gravitational force would be an additional component to the known standard gravitational constant G. The additional component, probably, would only manifest itself when matter achieves very high densities like when it is in the plasma state.

In Chapter XIII we'll return to this concept because of its impact on the development of the Universe.
Our theory proposed years ago the idea that dark matter was made up by large neutrinos concentrations. Lately astrophysicists have come to the same conclusion.

(1) This concept is expanded in Chapter XVI.

(2) For simplicity purpose we show only the nucleus instead of the complete atom. The electron's contribution to the force of gravity is about 2000 times smaller.

Chapter XII

The Dark Force or Dark Energy

NEUTRINO CASE No. 10

In this chapter we will explore what the astrophysicists now call "the dark force." First we will explain how this phenomenon manifests itself, and then we'll explain it according to our theory.

According to the standard view of cosmology, the once infinitesimal Universe has ballooned in volume ever since the original Big Bang, but the gravitational tug of all the matter in the cosmos has gradually slowed the expansion.

In 1998, however, scientists reported that a group of distant supernovas were dimmer—and therefore farther from Earth—than the standard theory would indicate. It was as if, in the billions or so years it took for the light from these exploded stars to arrive to Earth, the distance between the stars and our planet had stretched out more than expected. This would mean that the cosmic expansion has somehow sped up, not slowed down. Recent evidence has only confirmed this bizarre result.

In 1929, Edwin Hubble discovered that distant galaxies are

fleeing from each other as if the entire Universe is swelling in size. Ever since, astronomers have been hoping to answer one question: Will the expansion of the Universe, slowed by gravity, go on forever, or will the cosmos eventually collapse into a "Big Crunch"? Despite decades of efforts and countless studies devoted to the ballooning of the Universe, the recent findings stunned astronomers. Few suspected they had been asking the wrong question.

For the last 70 years, they've been trying to measure the rate at which the Universe slows down. When they finally did it, they found out it was speeding up. An accelerated expansion would seem to contradict all common sense. If we throw a ball into the sky, after it reaches a certain height, it will come down. Now imagine throwing another ball up and finding that, instead of falling back down, it somehow keeps moving up faster and faster. For that to happen there has to be a force pushing upward on the ball that is strong enough to overcome the gravitational pull downward. Astronomers have come to believe that this force is stretching the very fabric of the Universe.

To explain what we have been discussing here, first we have to develop in our minds a picture of what the visible Universe looks like. To help us in our task, we introduce Figure 12.1.This figure illustrates a wedge of the visible Universe from the Big Bang to the present day.

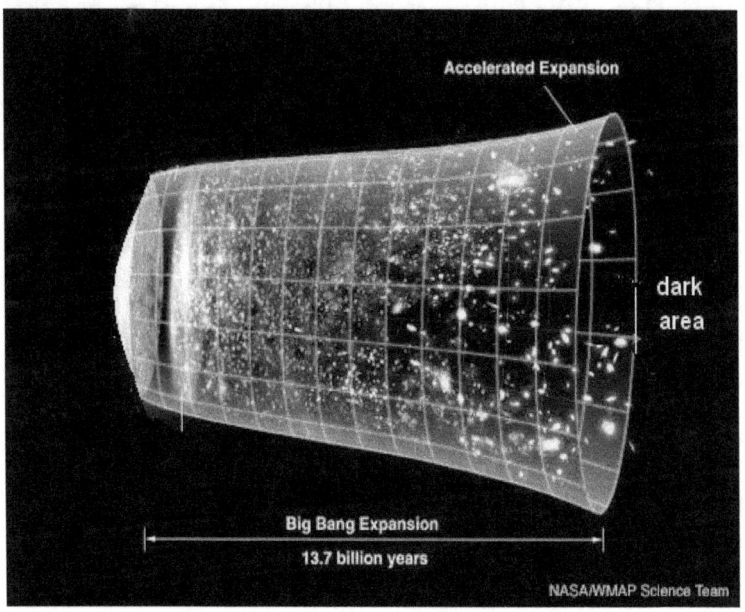

Figure 12.1-The Universe (NASA)

The gravitational field is indicated by the grid. In this picture the gravity lines stop where the visible Universe ends because we assume that no matter exists in the area marked "dark area" (only on the right side of the cone shape). Particles that do not generate light could exist in the dark area and generate their own gravity field. If the concentration of these particles in the dark area increases with time, then galaxies at the periphery of the Universe will move faster and faster toward the outside.

At this point we can make two different assumptions: one assumption is that every particle in this Universe is stable and that all these particles were thrown outward after the Big Bang. In this case, then we would expect that the expansion of the Universe should be slowing down and the galaxies in the periphery should be decelerating. Making a different assumption, a certain amount of dark matter is present in the dark area of *Figure 12.1,* and the gravity lines in the illustration should be extended to indicate the presence of this dark matter. This dark matter is made of neutrinos.

If we can explain why neutrinos, more than any other particles, have the tendency to increase their population in the dark area, we could say that we have found a way to explain why the expansion of the Universe is accelerating instead of slowing down. If the free neutrinos were moving out of the visible Universe faster than the rest of the matter, they would have the tendency to stretch the very fabric of the Universe itself.

In our theory there is a certain ratio between matter (neutrinos with photons) and free neutrinos in the visible Universe that is a component of the universal gravitational constant G. We are going to call G_1 this new component due to the ratio between free neutrinos and visible matter. We call G_2 the known gravitational constant due to the visible matter-visible matter reciprocal attraction.

$$G = G_1 + G_2$$

For the rest of the Chapter we are only concerned with this new component G_1 of the universal gravitational constant.

$$G_1 = \text{free neutrinos} / \text{matter} = \text{constant}$$

If we could explain why this ratio is changing and actually decreasing in the center of the Universe and increasing on the outer edges, then we would have an explanation for the speeding up of the cosmic expansion.

We think that the dark force is caused by supernovas continuously exploding in the billions of galaxies throughout the Universe. During the supernova phenomenon, large quantities of particles are destroyed and transformed into energy. When the particles are destroyed, they separate into photons and neutrinos. Free photons travel toward the periphery of the Universe in the same direction of the neutrinos they were associated with in the original particle. As long as the photon and the neutrino travel together, the following equation stays constant

$$G_1 = \text{free neutrinos} \,/\, \text{matter} = \text{constant}$$

Figure 12.2- A Supernova (NASA)

because both the matter (photons) and the neutrinos are moving together out of the part of the Universe that was once occupied by the star that exploded into a supernova.

In this case, we can call photons "matter." The reason for this is based on the fact that, when a particle is destroyed, photons carry more than 99 percent of the original particle. When a particle is destroyed in the supernova, the photon and the neutrino travel in a parallel travel. This parallel travel is not likely to continue indefinitely because the **photon is much more easily absorbed by interstellar matter.** Even the Universe background noise can interact with the photons. The neutrinos instead, with their extremely low interaction rate, keep traveling without their companion photons toward the extreme boundary of the visible Universe and beyond—into the dark area that surrounds the Universe. At this point, the matter in our equation, which is represented by the photons, stays behind, while the neutrinos continue to travel, and the relationship is no longer a constant.

$$\mathbf{G_1 = free\ neutrinos\ /\ matter \neq constant}$$

This means that what we call the "gravitational constant" **G** is no longer a constant because its component $\mathbf{G_1}$ is not a constant. The neutrinos, when they reach the dark area of our Universe, increase their pull on the galaxies that are located on the outside ring of the Universe creating what we call **"the dark force."**

CHAPTER XIII
The Death and Birth
of Matter

NEUTRINO CASE No. 11

In this chapter we look at two different types of celestial black holes. The first type is formed by the gravitational collapse of stars that are at least three times the mass of the sun. These black holes have a mass bigger than neutron stars but are essentially made up with the same substance called plasma. Plasma is composed of nucleons of atoms that have lost all their electrons and has a density of about 10^{17} kg/m^3. Another type of black hole is created by the gravitational collapse of a body billions of times the sun mass and has a density of about 10^{150} kg/m^3. These black holes, according to our theory are made exclusively by neutrinos, and due to their density, they are so small in size to qualify as a singularity.

Before reading the following paragraphs, we have to remind ourselves that the size of a nuclear particle is inversely proportional to its mass. We have shown in Chapters IV and V that the electron radius is approximately 2000 times bigger than the proton radius, although its mass is 2000 times smaller. If we split the proton into many parts, each of the parts would be bigger than the original proton. There is nothing magical about this fact. Due to the nature of the electromagnetic fields that carry most of the particle energy or mass, the more massive particles are smaller than the less massive ones. More massive particles mean higher frequency electromagnetic field and smaller wavelength. The wavelength defines the size of the particle.

[83]

In this chapter we're going to compromise the very existence of the particle by placing it in a space with very high *free- neutrinos* density. We have to remember that splitting the particle into smaller masses is not a solution, because the new particles size would only increase. In other words the particle is not like a loaf of bread, that if it does not fit in the bread box, we can cut in halves and put one of the halves in the freezer. In our case, cutting the loaf of bread in halves, would create two halves whose size is double the size of the original loaf. Another thing to remember is the fact that in order for a star to become a black hole its mass has to be at least three solar masses. In this chapter we'll call these black holes "galactic black holes" to tell them apart from the small black holes that have been the major topic of this book.

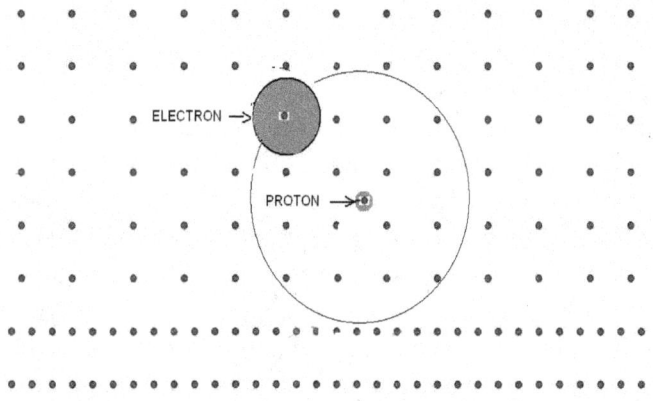

***Figure 13.1 - A Hydrogen atom in a gravity
field with a gradient***

In the *Figure 13.1*, one hydrogen atom is shown in a gravity field with a gradient. This figure shows an atom, with a proton in the center and an electron rotating around it. The picture is not drawn to scale otherwise the proton would be a point that is barely visible. The free neutrinos density increase toward the bottom of the picture, due to the effect of a gravitational field that becomes more intense in the downward direction. The hydrogen atom would experience a force pushing it toward the bottom. We could imagine the depicted gravitational field to be a portion of the gravitational field of a galactic black hole. The hydrogen atom will sink more and more in the denser and denser *free- neutrinos* fields.

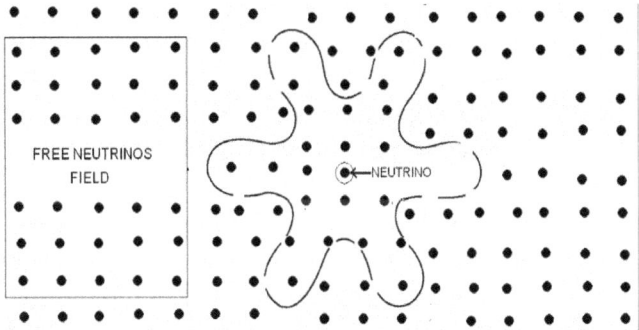

Figure 13.2 – An electron electromagnetic wave inside a free- neutrinos field

There will be a time when the density of the *free- neutrinos* is so great, that as shown in *Figure 13.2,* that the electromagnetic wave (*photon*) of the electron cannot freely rotate around the captured neutrino. We do not know what neutrino density is needed to disrupt the travel of the photon; this density is probably very high because of the regenerative properties of the electromagnetic waves.

When this critical density is reached inside a collapsing star, the electron electromagnetic field can no longer exist. The electron cannot split in smaller less massive particles, because such particles would be even bigger than the original electron, making the situation even worse.

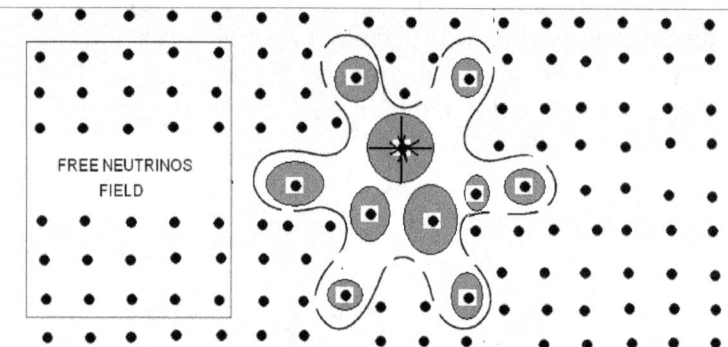

Figure 13.3 - The implosion of the electron electromagnetic wave in a galactic black hole

There is only one solution available; the electromagnetic field, in order to preserve its energy, has to collapse and create a different **entity** in the process. The only **entity** known to us that can fit this process is the neutrino. The implosion of the electron will generate many neutrinos, as shown in *Figure 13.3*. This implosion would generate neutrinos that probably do not have all the same mass. An equal amount of positive and negative spin neutrinos needs to be created since the polarized wave is a 0 spin photon. This transformation of photons into neutrinos could be an explanation why neutrinos are oscillating particles. Some of the properties of the photon are preserved in the neutrino.

All the atoms, that were part of the star before it collapsed, lose all the electrons and are transformed into plasma. In the new created black hole, all the matter is made of protons and neutrons and no electrons are allowed to exist. The new black hole will be about one mile in diameter when all the three sun masses have collapsed.
The following picture shows why the proton is capable of surviving the galactic black hole gravitational field.

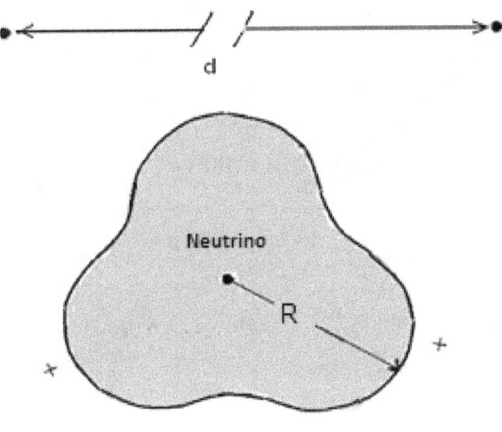

Distance between neutrinos d > 1000 R

Figure 13.4 - Proton in a galactic black hole gravitational field

Inside the galactic black hole, the proton, as shown in *Figure 13.4,* survives because its electromagnetic field, being about 2000 times smaller than the electron, is not disrupted by the density of the free neutrinos field. This picture shows an average distance **d,** among free neutrinos, resulting in a density comparable to the one in *Picture 13.2,* which shows that the proton electromagnetic field is not disrupted.

Now that we know the cause of the electron demise, we have to ask ourselves if there could be an event in the universe that could cause the implosion of the proton itself. Since we know that the electron demise was caused by the density of the free neutrinos field, if we could increase the density of the neutrinos to the point that the neutrino density could look like *Figure 13.5,* then a collapse of the proton would be possible. In order to achieve such density of neutrinos we need a *super massive extragalactic black hole.* We call it this way because is unlikely that such a black hole is present in our galaxy. A quick computation can be done by assuming that the free neutrinos field in this super black hole would have the distance between neutrinos 2000 times smaller than the one shown in *Figure 13.1.* To achieve this density we need a minimum of 30 billion solar masses.

$$\textbf{(3 solar masses)}(2000)^3 \approx 30\text{x } 10^9 \textbf{ (solar masses)}$$

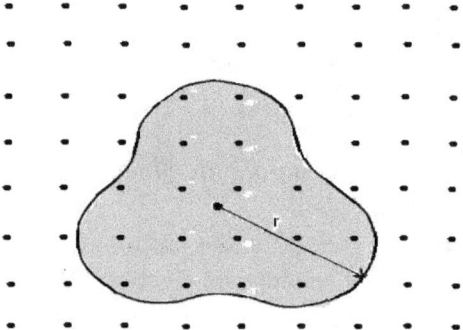

Figure13.5 - Proton in a 30 billion Sun masses
black hole gravitational field

This is an approximate computation and the actual number of solar masses needed could be few orders of magnitude greater. When the free neutrinos density of this super galactic black hole is achieved, then the electromagnetic field of our proton can no longer survive and starts imploding as shown in *Figure 13.6*. The energy of the collapsing electromagnetic fields can only be preserved by turning it into neutrinos, causing a great increase of the free neutrinos population. This condition is unstable, because it creates a positive feedback; the more protons and neutrons are destroyed, the more free neutrinos are created and the gravity field increases exponentially. Once the critical mass of the super galactic black hole is reached, all the particles, with the exception of the free neutrinos, will be destroyed in a very short time. The final size of this super galactic black hole is extremely small, so small to be classified as a singularity. We surmise that this process is **totally reversible**, this means that certain neutrinos created by the collapsing photons, can **transform themselves back into photons** if the gravitational field decreases.

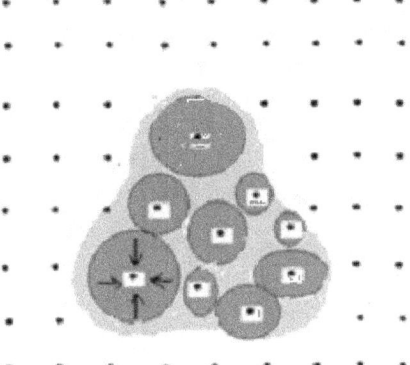

Figure 13.6 - The implosion of the proton

After the Big Bang, as the density of the *free- neutrinos* field decreased due to the expansion of the Universe, photons started to be created. Eventually these photons (*electromagnetic waves*) were captured and the first protons started appearing

Returning to our galactic super black hole, we do not know what kind of density is required for the neutrinos to start violating the Pauli Exclusion Principle, but this principle cannot be violated, and sooner or later the *galactic super black hole* would explode like in *Figure 13.7*.

Figure 13.7 - The explosion of a super galactic black hole

This event is similar to the Big Bang. From the explosion, the neutrinos start expanding at the speed of light. After a certain amount of time the black hole effect disappears and photons start to form. The neutrinos and the photons would eventually recombine creating a new generation of protons.

If more than 30 billion solar masses are needed to create a super galactic black hole, then it is very likely that this critical mass cannot be achieved by the mass present in any given galaxy, and this event would have taken place, only once, 13 billion years ago, at the time of the Big Bang.

Before the Big Bang, the Universe would have been just a singularity. This assumption is based on the fact that neutrinos are singularities and, no matter how many neutrinos were present in the pre Big Bang phase, this Universe would have been a singularity in itself. In this dark Universe only neutrinos were present. In this original singularity, there were no electromagnetic, gravitational, strong and weak nuclear forces, because for these forces to exist they need the interaction of neutrinos and photons, but photons were not present in the Universe before the Big Bang.

In this chapter, we were able to reach the conclusion that the Universe was originally a singularity. The fact that this same conclusion was achieved by many other scientists, using different mathematical methods, gives a lot more support to our theory.

Chapter XIV

The Oscillating Neutrino

<u>NEUTRINO CASE No. 12</u>

Recent observations give credit to the unusual fact that neutrinos can change their identity while traveling in space. Electron neutrinos can change into muon or tau neutrinos and vice versa.

In the last chapter we explained how it is possible in a strong gravity field for the electromagnetic waves to be transformed into neutrinos. In this chapter we look into the enigma of the oscillating neutrino. It appears that there is a pattern in the Universe, this pattern indicates that sooner or later everything has the tendency to be recycled. It is possible that the pre Big Bang Universe was the outcome of the collapse of something that existed in prior times. In this case the transformation of all the existing particles into neutrinos is even more credible since the neutrinos are the first particles to be ejected by the Big Bang. The Oscillating Neutrino Theory is even more captivating if we believe that neutrinos are generated by collapsing electromagnetic fields. This can explain the oscillating enigma since the neutrinos created would retain the oscillating property of the electromagnetic fields. They also could retain the capability of traveling at the speed of light, which is another corner stone of our theory.

To study this phenomenon we need to review in more depth the general black holes theory. We start by first computing the radius of the event horizon of a large black hole. To become a black hole, a body must contain a certain amount of mass. The gravitational field of a body is given by the following general formula:

$$g = G\,M\,/\,r^2$$

Where G is the gravitational constant, M is the mass of the body, and r is the distance from the center of the body to the location in space where the gravitational force g is computed. Next, we compute the intensity of the gravitational field that is required for the body to become a black hole. In this case, we set r equal to R, the radius of the event horizon. By definition, a black hole has a gravitational intensity at the event horizon that even a spaceship moving at the speed of light cannot escape. The following equation gives us the centrifugal acceleration of a spaceship moving at the speed of light C trying to escape the gravitational pull:

$$g \;=\; C^2/\,R$$

By equating the gravitational pull to the centrifugal force, we create the balanced condition that typifies the event horizon.

$$R \;=\; G\;M\;/\;C^2 = \textbf{event horizon radius}$$

At the event horizon, a photon can rotate around the singularity forever because the forces are in equilibrium.

From the above equation we can compute the mass needed for a body to become a black hole:

$$M = R \, C^2 \, / \, G$$

In *Figure 14.1*, the line from 0 to A defines the event horizon. A body of a certain mass must have a radius smaller than the computed event horizon R to become a black hole.

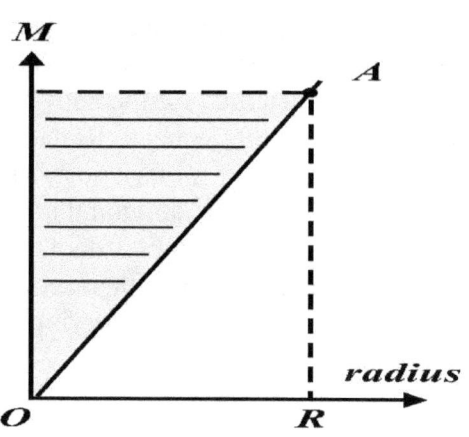

Figure 14.1 - The Black Hole Region

The shaded area is the black hole region. For the Earth to fall into the shaded region, it would have to shrink to the size of a small pea. This could never happen because, even if all the atoms of the Earth were stripped of their electrons, the Earth would still be the size of a large boulder. Normally, when stars implode, they become neutron stars with a density of 10^{16} Kg/m^3. This is the density of matter when deprived of all the electrons. Another way to call this condition is "plasma". This is the highest density that scientists believe that matter can achieve. Stars, when they implode, besides the need to meet this density, need a mass that is at least three times the mass of the Sun in order to become black holes.

In the following equations we compute the density ρ, or specific weight of a black hole:

$$\rho = M / \text{Volume} = R\ C^2\ /\ G\ 4\ \pi\ R^3\ /\ 3$$
$$\rho = 3\ C^2\ /4\pi\ G\ R^2$$

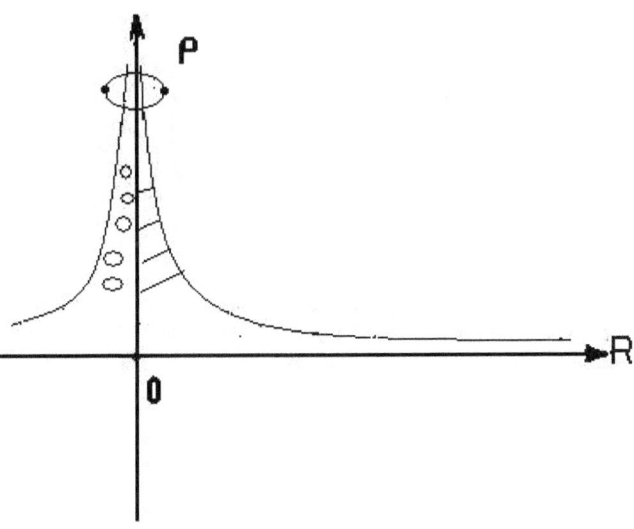

Figure 14.2
Specific Weight vs. Radius of Black Holes

Assuming that a density greater than Plasma is possible, the above computation tells us that the event horizon **R** of a black hole can be very small as long that the density ρ of its mass increases accordingly.

Figure 14.2 demonstrates that, for a body, the smaller the event horizon R, the greater is the specific weight requirement for achieving a black hole status. The specific weight increases exponentially as the radius contracts. This graphic is symmetrical with respect to the ρ axis, but the interpretation of the two regions, one right and the other left of the ρ axis, is completely different. The region to the right is the black hole region, and the region on the left, marked by circles, is the **wormhole** region. The wormhole region, since it is located in the negative region of the radius, is hard to interpret. Our interpretation is that anything in that region cannot be observed by an observer located in the positive region of space.

In the large celestial black holes, the atoms must lose all their electrons to achieve the black hole status. The Sun, even if contracted to the density of a neutron star, could never become a black hole. We need a star at least three times bigger than the Sun to make a black hole. The Sun is limited by the fact that even neutrons, as we covered in the main body of this book, are made of empty spaces.

Neutrinos are the only components of our Universe whose composition does not include magnetic waves. For the first time we encounter what appears to be a totally solid body. Neutrinos are not made by empty spaces and, therefore, there is no known limit to their density. From our formulas we can compute the density required for a neutrino to be a black hole. The required density is a mind bending figure of 10^{150} (1 followed by 150 zeros) times the density of **water.** Furthermore, we can compute the radius of the neutrino event horizon to be about 10^{-60} meters. It is surprising that such a small event horizon can bend the electromagnetic wave of a photon. Photons themselves are not well understood. We are familiar with the fact that electrons orbiting around the nucleus of an atom are able to capture photons with wavelengths in the order of 10^{-6} meters. The electron dimensions are about 10^{-12} meters.[1] This is already the case of a particle, many orders of magnitude smaller than the photon wavelength, that is capable of capturing the photon electromagnetic wave (photoelectric effect). We have to assume that, as long as that part of the electromagnetic wave touches the event horizon of a singularity, its own internal clock is temporarily stopped. In our particle model, part of the electromagnetic wave of the photon that surrounds the singularity is missing. This part is missing because, when the wave touches the event horizon, its internal clock stops. Although the event horizon radius is extremely small, it still has the capability to bend the wave into a circle. If the radius of a neutrino is 10^{-60} meters then the cross section would be 10^{-120} meters^2. This means that if we spread on a surface the 10^{85} neutrinos, the number we computed in Chapter III forming the total mass of the Universe just after the Big Bang, we would cover an area the size of the cross section of a proton. This would indicate that the direct collision of neutrinos and neutrinos is just about impossible. Some neutrinos decay whenever exposed to the change of a potential energy field. The potential field can be a gravitational field or the field of an electromagnetic wave. The latter decay happens whenever the neutrino interacts with matter; some of these interactions are observed in special laboratories on Earth.

Chapter XV

Ultra High Energy Cosmic Rays

NEUTRINO CASE No. 13

In 1962, the first observation of a cosmic ray with the energy of 15 joules was made by John Linsley and Livio Scarsi in an experiment in New Mexico. This finding was very surprising because nobody until then thought that it was possible for a cosmic ray to carry the energy of a baseball travelling at the speed of 60 mph. Fifty years later, there is still no good explanation for the origin of such rare [1] but very powerful rays consisting exclusively of protons. We believe that our theory has an explanation for such phenomena. In Chapter XIII, we discussed how, in a gravity field generated by a black hole greater than 50 billion sun masses, the proton would be destroyed and its electromagnetic field transformed into neutrinos. During this process, the gigantic black hole starts collapsing and reaches a size much smaller than a standard black hole. This happens because standard black holes [2] allow the survival of the protons and neutrons and therefore they retain a considerable larger size in spite of the smaller mass. The Universe before the Big Bang could have reached its incredible small size going though a similar process. But after the Big Bang, a new and opposite process was started and the neutrinos were changed back into electromagnetic fields. The first particles to be expelled by the explosion were neutrinos travelling at the speed of light. It took about ten years for the expanding Universe to exit the black hole event horizon; only then some of the neutrinos, through a reverse process, started decaying into photons. The original electromagnetic field of the proton had been compressed like a spring into a neutrino by the tremendous gravity of the pre Big Bang Universe. The amount of energy stored in these compressed photons had to be enormous because they were exposed to a gravity field greater than 10^{20} g. Then, some neutrinos started decaying, and photons were released from their compressed state. These photons started travelling at the speed of light among the remaining neutrinos. At this point some of the photons (gamma rays) were captured by some of the remaining neutrinos and formed the first protons. These protons for relativistic reasons could not travel at the speed of light but started travelling so close to the speed of light that, in a year race with light, they would have fallen behind only a few nanometers. We think that these very rare ultra high energy cosmic rays are just a leftover of the original protons compressed electromagnetic field produced by the pre Big Bang Universe or by huge black holes that eventually became unstable. The general population of cosmic rays is made by 90% protons and 1% electrons. The fact that the electrons carry considerable less kinetic energy appears to agree with our theory. After a correction for the difference in mass, the ultra high energy electrons carry about a million time less energy than the protons. It looks like the Universe is not capable of accelerating electrons as much as protons. In Chapter XIII, we described how the electron electromagnetic field would have been compressed into a neutrino by a black hole with only 3 sun masses. Instead a black hole needs at least 30 billion sun masses in order to compress the proton. Therefore, in the case of the electron, due to the weaker gravity field, the resulting compression energy stored inside the neutrino should be considerably less. Our theory can give no explanation for the fact that, in the composition of the cosmic rays, the protons are considerably more numerous than the electrons.

Further evidence that neutrinos are just a compressed form of the electromagnetic waves and eventually they decay, can be found in the following analysis. In Chapter XIV, we computed the neutrino cross section to be about 10^{-120} m^2 and if we converted all the mass of the present universe in neutrino masses we would get 10^{85} neutrinos. From the above figures, we could fit all the 10^{85} neutrinos cross section into the cross section of a proton and therefore deduce that interactions among neutrinos are extremely rare.

To clarify this concept we can use the following example: if we fit all 10^{85} neutrinos inside an old style keyhole and then we shoot through the keyhole the same quantity of neutrinos, it would be very unlikely that any direct neutrino-neutrino collision would happen. It is amazing but, if you could pass all the mass of the Universe through it, it still would be unlikely to observe a collision. A collision between galaxies can be an analogy. In this case, when two galaxies collide, a direct impact of their individual stars is extremely remote. Basically, according to our theory, it is almost impossible for the neutrinos crossing the Earth to collide directly with its matter. We believe that, due to changes in the Earth gravity and magnetic fields or due to the electromagnetic fields of the individual atoms in its vicinity, the neutrino sometimes decays or causes other particles to decay. These decays are rare, as shown by the low level of detection in various scientific facilities scattered around the globe, but not uncommon.

We find of interest that a gamma ray carrying the energy equivalent of a proton would have a dimension of about 10^{-15} m, but when this gamma ray is compressed into a neutrino, it acquires dimensions of about 10^{-60} m. This 10^{-45} magnitude difference is almost equal to the product of the gravitational constant **G** and the Planck constant **h**. Considering that this is an approximation over many orders of magnitudes, this equation is quite interesting.

Photon $_{size}$ x h x G $\approx$ Neutrino $_{size}$

This equation includes some of the most important constants of the Universe. The photon electromagnetic field includes the magnetic permeability μ and the dielectric constant ε and both of these constants are related to the speed of light **C**.

$$C = (1 / \mu \, \varepsilon)^{1/2}$$

Therefore, the equation of the **neutrino - photon** includes all the constants of the Universe. All these constants were created after the Big Bang.

Staying within the neutrino decay subject, we believe that the neutrino decay offers a solution to one of the problems we encountered in developing a logical explanation for our theory. In Chapter IX we said that the neutrinos are the only particles able to travel at the speed of light. They can travel at the speed of light because neutrinos are not

[101]

affected by the relativistic problem of the other particles whose mass becomes infinite at that speed. The neutrinos are just a different form in which the electromagnetic waves can manifest themselves and therefore are capable of travel at the speed of light. Unfortunately this property of the neutrinos presents a new problem. While all the other particles can be accelerated to acquire any kind of energy, the neutrino mass stays constant during accelerations and the maximum energy possible is the rest mass multiplied by the square of the speed of light. The problem is that the formula $E = mc^2$ limits the neutrino energy to a level well below the observed levels. However, neutrino decay allows the neutrino to acquire a huge amount of energy as explained in the previous paragraphs.

(1) These Ultra High Energy Cosmic Rays have the frequency of one every 100 year per square kilometer

(2) A 3 sun masses black hole has a diameter of about a mile. A 30×10^9 sun masses black hole can easily fit inside a Planck Length. See Chapter III.

Chapter XVI

The Clocks

NEUTRINO CASE No. 14

This chapter covers the expansion of the Universe starting with the Big Bang and explains the direct relationship between the **time** that elapsed from the Big Bang and the **evolution** of the nuclear particles. According to most astrophysicists, at the very beginning, just after the original explosion, there were only neutrinos. The expanding Universe, for about ten years, retained all the properties that are present inside the event horizon of a black hole. This means that time did not exist and the neutrinos kept expanding, for a while, in a timeless Universe. We know from the last chapter, that in a gravitational field **increasing** in magnitude, the protons are the last particles to be destroyed, therefore we can assume, that in a field **decreasing** in magnitude, they would be the first particle to be created. In this original expanding Universe, as the gravitational force decreased, the black hole effect disappeared. As soon as the expanding Universe event horizon vanished, some of the neutrinos decayed into **photons** (*this is the reverse process explained in the previous chapter*). Some of the remaining neutrinos captured these photons, and formed the first **protons**. Later on, with ever decreasing gravitational forces, electrons were created by other neutrinos capturing less energetic photons.

These captured photons, having a precise frequency of their own, were the first oscillators to populate the Cosmos. Just a few thousands of years after the Big Bang, the Universe developed not only time but trillions of trillions of oscillators that we'll call **clocks** for reasons that we are going to explain.

As soon as time appears so do the **clocks.** It is obvious that a clock without the existence of time makes no sense, but in order to keep track of time we need clocks and this is exactly what happened. To expand this concept we need to review what is our meaning of a clock.
 Old mechanical clocks have a mechanical oscillator like the pendulum and from the mechanical pendulum they derive through gears a rotational motion that moves every second the clock hand. Most modern wrist watches have an oscillator made by a quartz crystal and an electronic circuit that moves every second the watch second hand. All the clocks in order to keep track of time need an oscillator. The particles that have been described in the previous chapters all have photons as part of their structure and photons themselves are oscillators. The different particles are only differentiated by the different frequency of the photons. These particles have oscillators in the same way that a pendulum clock has an oscillator. Since the only essential part of a clock is the oscillator, we take the freedom to call nuclear particles, like electrons and protons, *clocks*. Hypothetically, a wrist watch could be made using a proton as an oscillator. We say hypothetically, because in order to observe the **proton oscillator** we need to do what is always needed for observing, to bombard the proton oscillator with other photons. This process would destroy any meaningful measurement of our attempt of keeping track of time.

The reason why we introduce this notion in this chapter is because the nuclear particle is not just a simple clock, but a clock that can tell when it is being accelerated or when it is being pulled in a gravitational field or in any other type of potential energy field.

Gravitation and kinetic energy

On Earth, when we shoot a bullet from a gun, the bullet will eventually land on the ground. If we shoot a second bullet with more kinetic energy, the bullet will land farther away from the gun. If it were possible to give a third bullet enough kinetic energy, it will eventually get into orbit around the Earth. In this case enough kinetic energy was given to the third bullet to overcome the potential energy that kept the bullet bound to Earth. In other words, we neutralized the pull of the Earth gravitational field. We know that inside the bullets there are billions and billions of nuclear particles and therefore there are billions and billions of clocks. If we compare the clocks of the first and third bullet, their oscillators have a different frequency which means there is a discrepancy in their time keeping. The third bullet, while into orbit, is retaining permanently part of the kinetic energy from the gun and this event is marked by the higher frequency of the particles clock. In order for the nuclear particles of the third bullet to find a new equilibrium they have to station themselves into a less powerful gravitational field of a terrestrial orbit. All the nuclear particles that form the fabric of an object, such as protons, neutrons and electrons, experience the changing of their oscillator frequencies, when exposed to outside forces. The final result of applying an outside force is the change of their kinetic energy or the change of their potential energy. These changes are marked permanently in the clocks of the nuclear particles. It has been experimentally proven using accurate atomic clocks that time is affected by the gravitational field. It is indeed miraculous that every nuclear particle is not just a piece of dead matter but essentially a precise clock. We can say that the nuclear particle, the building block of the Universe, is already a complex entity and its structure is designed for easily building more complicated forms of life.

The String Theory

We had to use a gun to apply kinetic energy to a bullet but if we had to accelerate a single electron of an atom the technique to be used would be different. We would target the atom with photons and when one of the atom electrons absorbs a photon it will acquire kinetic energy and the particle internal clock will increase its frequency. The electron will find then a new equilibrium position in a new peripheral orbit where the increase in kinetic energy will be compensated by a decrease in potential energy of the electric field generated by the nucleus.

The same reasoning that was applied to the electrical field can be applied to the potential energy of a static magnetic field. We can conclude that the internal oscillator of a particle can detect movement in space due to a change of its coordinates in any given potential energy field. To these three variables representing the three coordinates in space, we can add other variables, such as the one connected to time, the one connected to the gravity field, the one connected to the electric field, and the one connected to the magnetic field. In total, our particle *oscillator* can detect

seven different variables. Some of these variables have a positive and negative region; if these regions are taken in consideration they further increase the number of variables. Therefore, the nuclear particles exist in a world of at least eleven different dimensions. Some people say it is impossible to perceive more than three dimensions, but what is perceived through our eyes is already in four dimensions because the view changes with time.
If our eyes could detect a gravitational field or an electrical field, then the number of dimensions in our perception would increase.

What was presented in this chapter, we believe, can be related to a theory physicists call **String Theory**. We perceive the String Theory, with all its abstruse formulations, like something that hangs on a limb disconnected from any other universal theory. Our theory instead is very inclusive and makes our version of the String Theory an integral part of a new nuclear particle theory.

Chapter XVII

The Photoelectric Effect

NEUTRINO CASE No. 15

When an atom looses one of its electrons due to the interaction with a photon, this effect is called the Photoelectric Effect. In this case, the freed electron totally absorbs the photon making it disappear and this is consistent with the behavior of a black hole. We like to dwell on this fact because in the Universe the only bodies capable of capturing photons are black holes. The black hole gravitational pull is such that not even something moving at the speed of light like the photon can escape. In this case the electron is definitely acting as a black hole. When the photoelectric effect was discovered, the black hole theory had not yet been proven and this fact can explain how the scientists made no attempts to correlate the two phenomena. It is amazing that the scientific community never revisited these phenomena in view of the present discoveries on the black holes.

When an electron captures a photon its kinetic energy increases to the point that the electric field of the atom nucleus cannot any longer keep the electron in its orbit and therefore the electron escapes. As it was discussed in the previous chapter, the frequency of the internal clock of a particle increases when the kinetic energy increases. In the Photoelectric effect the frequency of the electron internal clock increases after the absorption of the energy of the photon. The incident photon energy is equal to **hf** where the

f stands for the frequency of its electromagnetic field and **h** is the Planck constant. The electron, in order to escape will have to overcome the potential energy of the nucleus electric field, therefore only the partial energy ΔE will be transferred to the freed electron.

$$\Delta E = h \times (f - f_0)$$

Where hf_0 is the energy required to overcome the potential field of the nucleus. The following expression can be used to specify the actual energy transferred to the electron.

1.0) $\Delta E = h \times (\Delta f)$

Now we can write the new frequency of the internal clock of the electron that escaped due to the Photoelectric effect.

$$E = h \times (\Delta f + v)$$

In the above equation **v** is the original frequency of the electron internal clock. Like the internal clock of a particle leaving a gravitational field, the internal clock of a particle leaving an electric field increases its frequency by the amount Δf.

From equation 1.0 we can derive the following equation.

$$\Delta E \times \Delta T = h$$

From the above equation is possible to derive the Heisenberg indetermination principle equation and, with further derivations, the Schrodinger equation. [1] The Schrodinger equation when applied to the Hydrogen atom gives surprisingly accurate results. But this same equation already needs to be modified when applied to external orbits of electrons in heavier elements.

Although quantum mechanics gives excellent results in the study of atoms and molecules, we think that it cannot be utilized to study the inside of nuclear particles.

On the contrary, we have achieved surprising excellent results applying classical electromagnetic theory to our nuclear particle model in Chapters IV, X and XXI.

1) This last derivation was explained in the author's previous book "Travel into a Small Universe" Chapter XV.

Chapter XVIII

The Bullet Cluster

NEUTRINO CASE No. 16

Evidence of the neutrino acting as a small black hole can be found looking at faraway places in the Universe, and this is our task in this chapter. We have to remember that, in the last few years, astrophysicists were able to show the existence of black matter by looking into colliding galaxies and they have determined that black matter is made of neutrinos. *Figure 18.1* is the photograph of two colliding galaxies of the so called Bullet Cluster. The Bullet Cluster photograph on the book cover shows in blue color the dark matter and in red color the visible matter. The galaxies appear to be almost transparent because this picture is the result of combining photographs taken using different types of sensors. The collision was caused by the reciprocal attraction generated by the gravitational force. This force creates a separation of the galaxies' dark matter and invisible matter. Every galaxy contains a quantity of dark matter that is four or five times the quantity of visible matter.

According to our theory, during a collision, the **dark matter** (*free neutrinos*) of one of the galaxies experiences an attractive force due only to the **visible matter** of the other galaxy. The **visible matter**, instead, experiences an attractive force due to both the **dark and the visible matter** of the other galaxy. Therefore we believe that the dark matter, being subject to a lesser gravitational force, should trail visible matter in these circumstances. The blue

[111]

dark matter of both galaxies on the far left and on the far right of the picture clearly trails the red visible matter in the collision. The red visible matter of the two galaxies is merged to form almost a single galaxy.

Figure 18.1- Bullet Cluster Galaxy

In the following picture (*Figure 18.2*) we show the two galaxies that form the Bullet Cluster **before the collision**. For simplification purpose we put three protons representing the visible matter and three free neutrinos representing the dark matter in each galaxy. We have a total of six nuclear particles in each galaxy. All the six particles from galaxy A attract the three particles of visible matter in galaxy B, because in galaxy A we have six neutrinos attracting with their black hole effect the three photons belonging to the protons in galaxy B. At the same time, the dark matter of galaxy B feels only the pull of the three photons in galaxy A because free neutrinos do not attract each other. Of course, the same can be said of the attraction that galaxy B has on galaxy A. Therefore the visible matter is subject to a greater acceleration then the dark matter.

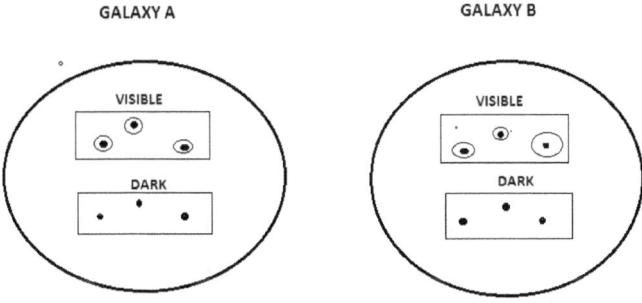

Figure 18.2 –Initial Position of Colliding Galaxies

In *Figure 18.3* we show the final result with the visible matter of the two galaxies merging in the red region and the dark matter left behind in the two blue regions. This is an approximate simulation of the Bullet Cluster scenario in *Figure 18.1*, where we see the actual separation of the dark matter from the visible matter as if the two galaxies went through the action of a sifter. The fact that not all the collisions offer the same scenario could have the following reason. The collision of two galaxies shows the dark matter trailing the visible matter only when the galaxies axis is perpendicular to the plane of the collision.

In the case where two colliding galaxies move along a trajectory that coincides with the axis of the galaxies, then the dark matter cloud is not left behind.

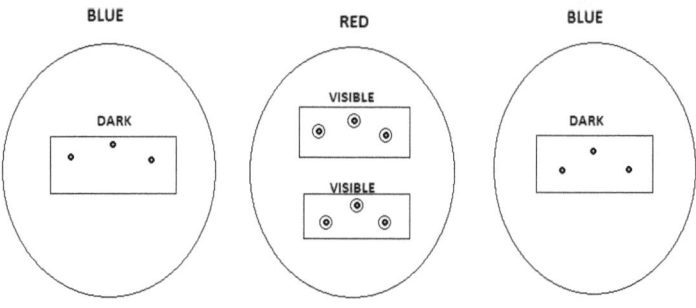

Figure 18.3 – Final Position of Colliding Galaxies

It will not be left behind because in this case the magnetic field effect on the dark matter created by the gigantic black hole in the center of the galaxy should be able to diminish the gravitational effect described in the Bullet Cluster collision. The next chapter will cover this effect in details.

Chapter XIX

The Paramagnetic Effect

NEUTRINO CASE No. 17

In this chapter we look at different galaxies in the Universe and try to establish a link between our nuclear particle model and the rest of the universe. First we have to recollect that, independently from its frequency, a photon captured in the event horizon of a black hole creates a charge no matter the size of the black hole. A captured photon manifests itself as a rotating charge or electrical current due to the warping of its electromagnetic field. These phenomena are independent of the size of the black hole. In neutrinos or in gigantic black holes, the captured photons generate an electric current. After this clarification, we can introduce in the following figure how presently the actual model of a galaxy is perceived.

In this *Figure 19.1*, we show a galaxy with a gigantic black hole in the middle whose mass is millions of solar masses. Billions of stars surround the black hole. The photons captured in the event horizon of the black hole generate a huge electrical current loop. An extremely powerful magnetic field is created by the current. The magnetic field accelerates a powerful jet of charged particles that are spouted out from the center of the galaxy. This effect is clearly visible in the composite photograph of *Figure 19.2* taken from the MPG/ESO telescope .This is a photograph of the Centaurus A galaxy and its jet of particles perpendicular to the plane of the galaxy exactly the same as *Figure 19.1*. This supports the validity of the model.

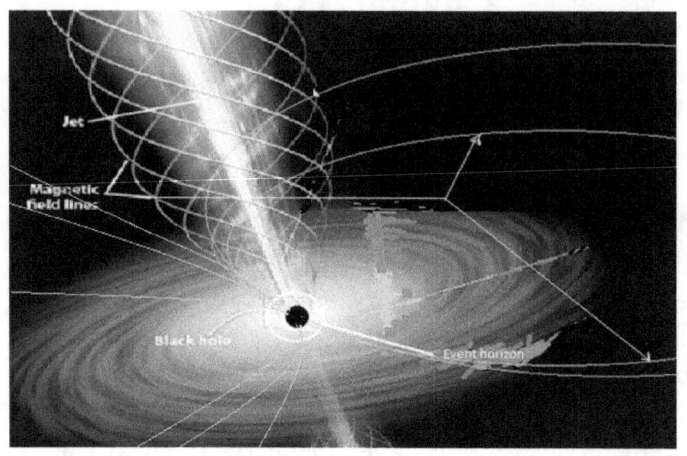

Figure 19.1 – The Galaxy Magnetic Field Model

Figure 19.2 – ESO- Centaurus A Galaxy

At this point we can go back to our nuclear particle model and make some further speculations on the nature of the neutrino. We think that the neutrino behaves like a very slightly paramagnetic material. This behavior will be apparent only under the influence of extremely strong magnetic fields. Only in this case, a neutrino can be affected like in the magnetic field generated by the photon in our electron model shown in *Figure 19.3*.

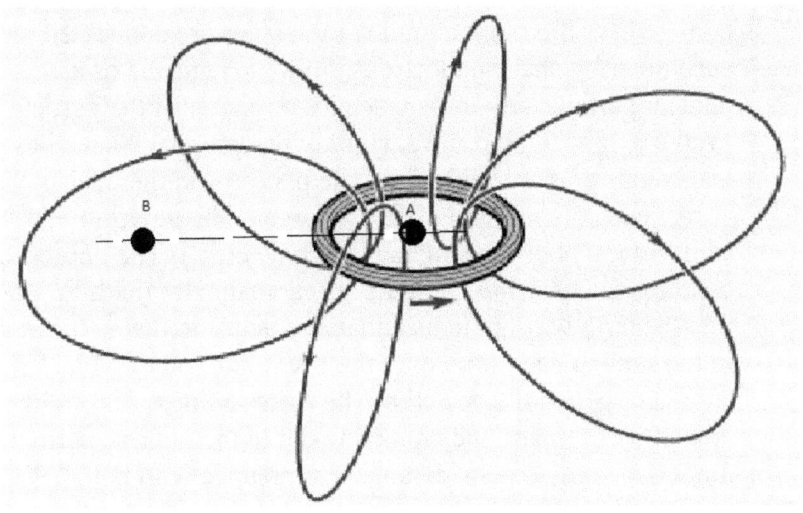

Figure 19.3- The Electron Nuclear Model

The electron model has a neutrino in the center and a ring around it representing the captured photon. The photon warped electromagnetic field produces the equivalent of a 20 Amps current moving around a circle with 10^{-12} meters diameter (see Appendix A). This condition creates a very strong magnetic field in the middle of the photon. Because

[117]

of this strong magnetic field, the neutrino finds the best equilibrium in the center of the photon where it acquires the maximum potential energy at position **A**. Other equilibrium positions can be found on the plane defined by the rotating photon. These locations are where the magnetic forces are null. Position **B** is one of the many locations meeting this last requirement. In the previous chapters we discussed how the neutrino captures a photon in the process of building a more complex particle. In this model there is a question to be answered. The question is: how in a solid body, where the more complex molecules are connected to each other by sharing electrons, does the neutrino of a nuclear particle stay in its place in the center of the particle even under acceleration? A slightly paramagnetic neutrino is the answer to our question. The photon and the neutrino attract each other using different forces. The neutrino captures the photon using is black hole effect. The photon captures the neutrino using its strong magnetic field. From this scenario we can deduce that any neutrino that is located in the same plane defined by the event horizon of a black hole is largely not affected by the surroundings.

When we compare the galaxy model to the nuclear particle model, we notice considerable resemblance with both having a black hole in the center and both having a strong magnetic field. We know that galaxies contain a large amount of dark matter or neutrinos and according to our theory this cloud of dark matter should reside in the flat plane defined by the black hole event horizon. This is where the magnetic forces are null and clouds of dark matter accumulate over millions of years of build up. This is where the stars are eventually born. Therefore the stars that are formed in these clouds of dark matter should reside more or less in a flat plane perpendicular to the magnetic field. Though this field is not strong when compared to an electron magnetic field, its effect accumulates over millions of years. The following *Figure 19.4* shows the Sombrero galaxy as photographed by the Hubble telescope. The flatness of this galaxy appears to validate our theory.

Figure 19.4- The Sombrero Galaxy

According to our theory, dark matter resides on the plane of the galaxies where it cannot be acted upon by any magnetic forces generated by the gigantic black hole. Only the inertia keeps the dark matter traveling with the companion galaxy, probably from the time it was formed after the first generation of stars disappeared. At this point we can go back to what happened after the Big Bang.

AFTER THE BIG BANG

After the Big Bang, the neutrinos were scattered at the speed of light in all directions. The density of matter was such that, only after ten years of expanding, this young universe was able to exit the black hole effect. At this point, some of the neutrinos started decaying into photons. These photons were captured by other neutrinos forming the first protons. The hydrogen nucleus was finally born. When a few hundred thousand years had elapsed from the original explosion, the gravitational force had decreased enough to enable the formation of the first electrons. With the presence of the electron, the creation of the hydrogen atom became possible. Millions of years after the Big Bang, we have large clouds of this hydrogen gas. Because of the gravitational force of attraction, these large clouds of gas started gathering and created the first generations of stars. The stars of the first generation were thousands of times bigger than the present average star, and because of the nature of massive stars, their life span was much shorter. In every core of these first generation stars there was a raging thermonuclear furnace that, by fusing together the hydrogen atoms started generating the heavier elements. Elements like helium, carbon and oxygen were formed. When this first generation of stars ran out of hydrogen it started producing energy with the fusion of heavier elements like carbon. The last and heaviest element to be formed in this fusion furnace was the iron atom. When the star ran out of all the fusible elements, exhausting all the available fuel, it started cooling down. The star core started collapsing and created a very unstable condition. At this point this first generation star, thousands of times the mass of the Sun, exploded in a gigantic supernova. In this monstrous supernova explosion, temporarily, there was the condition required to fuse atoms like iron into more heavy elements such as silver, gold and uranium. At this time a cloud rich of elements capable of combining themselves in many different forms was scattered into space. During the supernova explosion part of the original core remained in place and compressed itself into the first generation of black holes. When the first generation stars exploded there were two important developments that eventually created life. The first is the scattering of heavy elements throughout the cosmos, elements like the ones that became an active part of life on earth. The second is the formation of the first generation of gigantic black holes that eventually became the center of the new forming galaxies.

Chapter XX

Antimatter

NEUTRINO CASE No. 18

The quantity of antimatter present in the Universe is unknown; some of the visible galaxies could be made entirely of antimatter. We decided to cover the antimatter subject in this chapter because it supports our particle model theory. The fact that antimatter is easily produced in a laboratory setting is totally predicted by our model. In our theory, 50 percent of the particles produced are expected to be made of antimatter.

When a photon traveling in space is captured by a singularity, the particle that is produced has just as much chance to become negatively charged as it does to become positively charged. If a particle is created with a negative outside field, matter is formed; if the outside field is positive, antimatter is formed. We call them electrons or positrons depending on the outcome. Therefore we have to consider that antimatter is just a small variation of our basic particle model. The antimatter equivalent of an electron is a positron. The antimatter equivalent of a proton is an anti proton. The only difference between these particles is the charge, the electron being negative and the positron being positive; the opposite is true in the proton case. The difference in the charge is achieved by the electromagnetic field that is wrapped around the singularity; this field is 180 degrees out of phase or of opposite polarity in the antimatter particle configuration. In the *Figures 20.1 and*

[122]

20.2, we depict on the left a particle with its rotating photon, and on the right the same particle with the equivalent photon electromagnetic field. We selected a six-cycle wave because it is easier to visualize, though no stable particle has such configuration. Both *Figure 20.1* and *Figure 20.2* depict a particle and its antimatter equivalent. We do not specify which one is the matter and which is the antimatter because either pictures can represent matter or antimatter depending on the particle being observed. The only difference between the two illustrations is the standing electromagnetic wave, which in *Figure 20.1,* has a positive external field or positive charge, and in *Figure 20.2*, it has a negative external field or negative charge. In the case of the electron, *Figure 20.2* is what we call matter and *Figure 20.1* is the positron or antimatter. In the case of the proton and antiproton, the reverse of the illustrations would apply.

There is nothing superior or desirable about matter when we compare it to antimatter. Our Universe could be completely built with antimatter and there would be no difference in its appearance or inner workings. It is likely that, from the beginning, or Big Bang, what we call matter particles had a slight numerical advantage over the antimatter particles. Due to the fact that these two types of matter cannot coexist, this numerical advantage, over the eons, became ever greater, wiping out most of the existing antimatter. At this time, scientists believe that a cloud of antimatter exists in our own galaxy some 4000 light years from our solar system. All the stable particles that we observe in the vicinity of our solar system are made of matter instead of antimatter. It is possible that, during a new Big Bang, a new universe could be completely built with antimatter, and the antimatter people in it would live just as comfortably as we do. The two different types of universe cannot coexist, however, because of the most important characteristic of matter and antimatter: when a particle and its anti particle meet, they are both annihilated. The mass of both the matter and antimatter is totally converted into energy. This energy consists of high-energy photons—the same as the ones shown in our particle model. The amount of energy that is generated is given by the formula: $E = mc^2$

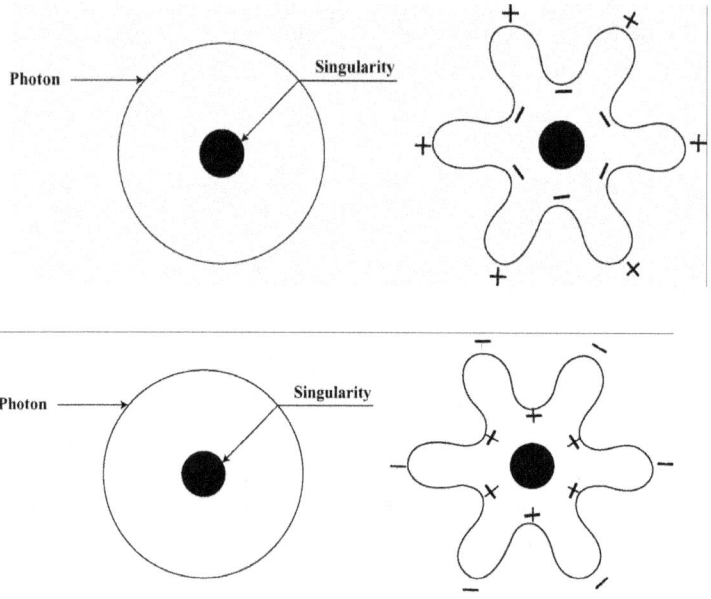

Figures 20.1- 20.2 - Matter and Antimatter are defined by the opposite polarization of fields

This total conversion of matter and antimatter into energy makes antimatter the ultimate source of energy. The combining of one gram of matter and antimatter would create an annihilation that would produce the energy of a

20-kiloton atomic bomb, the same size of the ones used at the end of World War II, where many pounds of plutonium were used. It is unlikely anyone will ever make an antimatter bomb because of the instability of the final product. Furthermore, because the energy is released mostly as gamma rays, an antimatter bomb would be only effective at destroying living organisms.

Antimatter is, first and foremost, matter. Whether positrons, antiprotons, or entire antimatter galaxies, antimatter should not be regarded as anything exceptionally different from what we call "matter." The fact that antimatter is easily produced in the laboratory concurs with our theory that 50 percent of the particles produced are antimatter. Though antimatter is easy to produce, it is difficult to store because any contact with matter must be avoided. Let's take a look at how antimatter is produced and stored. The most usual way to make antiprotons is to accelerate normal protons to high velocity in a proton synchrotron accelerator. At Fermilab (Fermi National Accelerator Laboratory) in the United States, the proton accelerator is operated at 120 GeV. These protons have over 120 times as much kinetic energy as their rest mass energy. They have enough energy to make 120 copies of themselves, or 60 pairs of protons and antiprotons. These high-energy protons are shot like bullets into a metal target. They collide head-on into the metal nuclei. Their kinetic energy is released at a single point in space in a tiny fireball. The kinetic energy turns into a spray of photons and pairs of particles and antiparticles. A small fraction of those pairs are protons and antiprotons. At Fermilab, the efficiency of this process is 4.7 percent. This means that for each of the 120 GeV initial protons, the average energy of the pairs of protons and antiprotons is 6 GeV. It is interesting to note that, for each proton an antiproton is generated, showing that in nature there is no bias toward matter or antimatter. Once the antimatter is generated, the biggest problem is storing it. Matter and antimatter have opposite electrical charges; therefore, they are strongly attracted to each other and, if left free to move, would mutually self annihilate in a fraction of a second. The researchers must capture the antimatter and put it into a bottle. Scientists in Switzerland have demonstrated a solution to this problem. Their container or "**bottle**" is an evacuated tube about two inches in diameter and bent into a ring 1000 feet in circumference. As the beam of anti proton goes around, it is directed and focused by a magnetic field to keep it from hitting the walls of the tube.

A Penning trap, which is considerably smaller, is usually considered a better bottle. The side walls of the Penning trap consist of a carefully machined metal ring about two inches in diameter. This trap is placed in a vacuum chamber inside the magnetic field of a superconducting magnet. The magnetic field runs along the axis of the donut-shaped trap, keeping the antiproton rotating in a circle. The walls of the trap are charged with a negative electric potential to keep the antiprotons away from the walls. If the magnetic field is strong enough, the antiprotons never touch the walls and are permanently trapped in a tiny circular orbit. In some experiments with the Penning trap, antiprotons have been kept in orbit for more than a month. At this point, the longevity of antimatter in captivity is dependent only on the level of the vacuum available inside the trap: the higher the vacuum, the less probability that the antimatter will collide with matter.

[126]

Chapter XXI

Theory of Relativity Revisited

NEUTRINO CASE No. 19

In this chapter we analyze how the travel of a light wave or photon is deviated when it is exposed to a gravitational field. We apply a new approach to a well-known problem solved by Einstein in his General Theory of Relativity. This problem, that concerns the deviation of the light of a star that reaches the Earth after it is affected by the gravitational field of the Sun, was originally solved using Riemannian math. The following calculations show that solutions identical to the ones in the General Theory of Relativity can be achieved using a different method. This new method is much simpler because it already incorporates the effect that the gravitational field has on time. There is no detriment done by incorporating this effect, since the relation between time and the gravitational field is already well established. This chapter gives a new understanding of how matter and gravitational fields interact with each other. Using our method, we can compute a deflection angle $\theta = $ **1.72"** of degree that matches the result of the General Theory of Relativity. This is the amount of deflection that a star ray experiences because of the gravitational field of the Sun while traveling to Earth.

We first start solving this problem by representing a polarized light wave moving along the Z-axis of a three-dimensional space. We represent the light wave with a Poynting vector:

$$\mathbf{Pz} = (\mathbf{E_x}) \ \mathbf{x} \ (\mathbf{H_y})$$

Where the Poynting vector is equal to the cross product of the electrical field E_x and the orthogonal magnetic field vector H_y. Both these fields can be written as:

$$\mathbf{Ex} = \mathbf{Exo} \ \mathbf{sin} \ \{\ \omega \ [t - z \ (\mu \ \varepsilon)^{1/2}] \ \}$$

$$\mathbf{Hy} = \mathbf{Hyo} \ \mathbf{sin} \ \{\ \omega \ [t - z \ (\mu \ \varepsilon)^{1/2}] \ \}$$

Both these vectors travel at the speed of light in their respective x and y directions for half of the wavelength and spring back to zero in the remaining half of the wavelength. From this we can deduce that the wave has a dimension in the x direction equal to:

$$\mathbf{X} = \lambda \ / \ \mathbf{2} \ \mathbf{sin} \ \{\omega \ [t - z \ (\mu \ \varepsilon)^{1/2}] \ \}$$

And the total displacement for a full cycle is:

$$\mathbf{X} = \lambda / 2 \ - \ \lambda / 2 = 0 \quad \textit{(Equation 21.1)}$$

The wave moves in the positive direction on the X-axis for one-half of the period and in the opposite direction for the other half of the period with no total displacement in the X direction.

We assume now that this same wave is moving in a gravitational field that has a gradient in the X direction and an orientation that affects only the electrical field **Ex** and has no effect on the magnetic field. We choose a polarized photon to simplify the math. A non polarized photon would give the same result, but would require also taking into account the effect of the magnetic field.

For the wave in *Figure21.1, Equation 21.1* is not true because:

$$\omega\, t1 > \omega\, t2$$

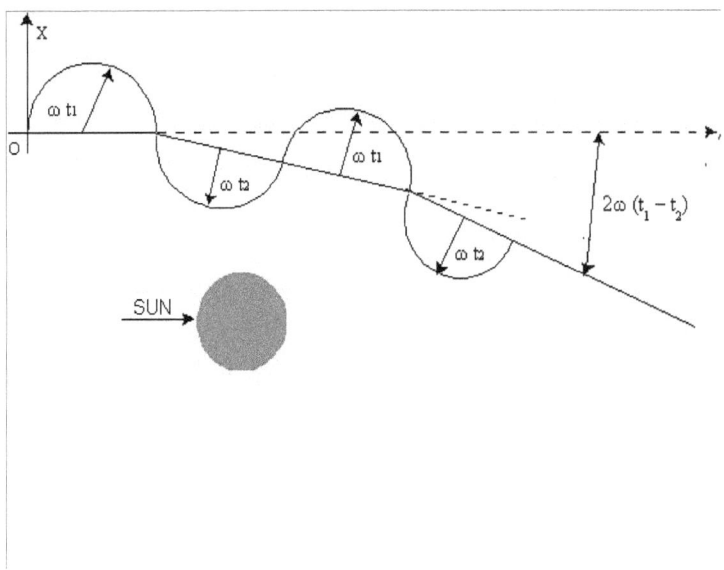

Figure21.1 - Time Warping due to the Sun's Gravity

This change happens because the negative wave, being closer to the sun, is subject to a stronger gravity field and, therefore, a slower time dimension, and for each cycle the wave is deviated by a very small fraction of an angle $\Delta\theta$:

$$\Delta\theta = \omega\,(t_1 - t_2) = \omega\,\Delta t$$

This is the deviation from the Z-axis for each cycle. For all the travel of the ray along the Z-axis we have a total deviation θ which is the summation of all the fractions of angle:

$$\theta = \Sigma\,\Delta\theta$$

This is the total angle of deviation of the light ray arriving on the Earth's surface.

The time warping can be computed using the following formula that has been experimentally proved:

$$\Delta t = t_0\,g\,\Delta r\,/\,C^2 \quad (\textit{Equation 21-1})$$

Where $t_0 = R\,/\,C$ is the time for the speed of light to travel from the center to the surface of the Sun, g is the gravitational acceleration due to the Sun, C the speed of light, and Δr is the sinusoidal change of the distance of the ray electromagnetic field from the center of the Sun. R is the radius of the sun.

$$\Delta\theta = \omega\,\Delta t = \omega\,t_0\,g\,\Delta r\,/\,C^2$$

The acceleration g due to the sun is:

$$g = G\,M\,/\,r^2$$

G is the gravitational constant. M is the mass of the sun, and r the distance from the center of the Sun. The **r**, the distance of the ray from the center of the Sun, is a function of both the X-axis and the Z-axis.

$$\Delta\theta \ = \omega\,\Delta\,t \ = \omega\,t_0\,G\,M\,\Delta\,r\,/\,r^2\,C^2 \quad \textit{(Equation 21.2)}$$

From *Figure 21.2* we can compute **r** :

$$r^2 \ = \ X^2 \ + \ Z^2$$

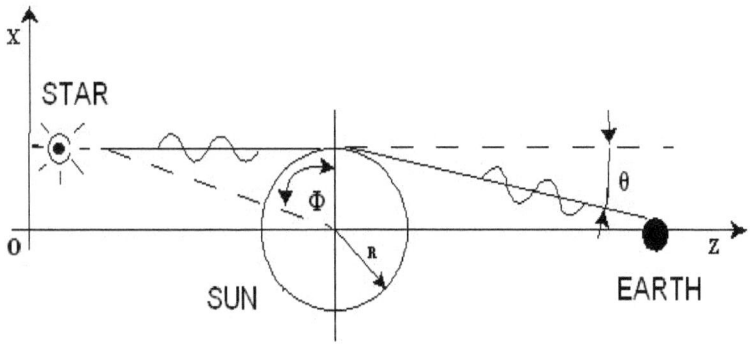

Figure21.2
Bending of the light of a star

      To simplify the computation, since the ray grazes the surface of the Sun, we call the radius of the Sun equal to 1.0; therefore we have:

$$r^2 \ = 1 + \ Z^2$$

As the light ray moves along the Z-axis, we have to compute for each cycle the angle of deviation due to the change in gravity between the semi-wave farther from the Sun and the semi-wave closer to the Sun. Looking back to *Equation 21.2* we determine that the Δ r changes in a sinusoidal mode:

$$\Delta r = \lambda/2 \quad \sin \{\omega \ [t - z \ (\mu \ \varepsilon)^{1/2}] \}$$

To simplify the computation, we replace the sinusoidal wave with a square wave as shown in *Figure 21.3*. The amplitude of the square wave has the average value Δ r.

$$\Delta r = \lambda / \pi \quad (\textit{Equation 21.2})$$

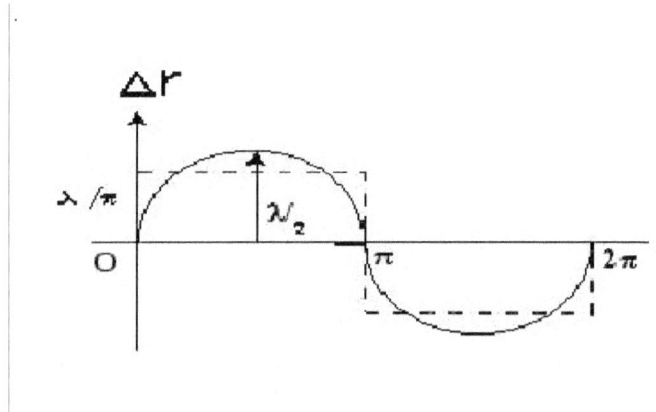

Figure 21.3- Distance Change From The Center
Of Sun
Therefore combining *Equation 21. 2 and Equation 21.1:*

$$\Delta t = t_0 \, g \, \Delta r / C^2 = t_0 g \, \lambda / \pi \, C^2$$

While the gravity field varies according to *Figure 21.2*:

$$g (z) = g \, [\, 1 / (\, 1+Z^2) \,]$$

By substituting for g(z), we find that:

$$\Delta t = t_0 \, g \, (z) \, 2 \, \lambda / \pi \, C^2 =$$

$$= t_0 \, g \, \lambda(\Phi) / \pi \, C^2(\, 1+Z^2) \qquad \textit{(Equation 21.3)}$$

We introduce $\lambda \, (\Phi)$ to take into account that the motion in the X direction is perpendicular to the gravity field only when crossing the axis of the Sun.

$$\lambda \, (\Phi) = \lambda \, \cos(\Phi)$$

Therefore, the effective wavelength is computed integrating Φ from $-\pi/2$ to $\pi/2$ and dividing over the interval π :

$$\lambda_{average} = \int \lambda \, \cos \, (\Phi) \, d\Phi \, / \, \pi = 2 \, \lambda / \pi$$

This result is the average value of $\lambda \, (\Phi)$. Inserting this value inside *Equation 21.3* we get:

$$\Delta t = = t_0 \, g \, 2 \, \lambda / \pi^2 \, C^2(\, 1+Z^2)$$

To find the total delay we compute:

$$\Delta t_{total} = \int t_0 \, [g \, 2 \, \lambda \, / \, \pi^2 \, C^2 \, (1+Z^2)] \, dz$$

We integrate over the limits (infinite to - infinite). An insignificant error is introduced using these limits since the distance of the Earth from the Sun is much greater than the Sun's radius. This is even truer for the distance that separates the Sun from the star.

$$\Delta t_{total} = 2\, t_0\, g\, \lambda\, /\, \pi\, C^2$$

$$\theta = \omega\, \Delta t_{total} = \omega\, 2\, t_0\, g\, \lambda\, /\, \pi\, C^2 = 2\, \pi\, f\, 2\, t_0\, g\, \lambda\, /\, \pi\, C^2$$

And replacing f with $C\, /\, \lambda$:

$$\theta = 4\, t_0\, g\, /\, C = 4\, G\, M\, /\, R\, C^2$$

Where G is the gravitational constant, M the mass of the Sun, and R is the radius of the Sun. This means that the angle of deviation is independent of the frequency of the light.

After we input the actual value of the above constants and convert the angle from radiants to seconds of degree, we obtain the total angle of deviation:

$$\theta = 1.72" \text{ of degree}$$

This angle of deviation is the same we find in the General Theory of Relativity. We can see from the above result that this new method of computing the effect of the gravitational field on photons is valid and simple. This chapter has proven that our theory of the New Particle Model, where photons paths are bent by neutrinos into a circular orbit, is credible.

[134]

For computation purpose the following constant were used:

$G = 6.67 * 10^{-11}$ **gravitational constant**

$M = 1.96 * 10^{30}$ **Kg**   **mass of the Sun**

$R = .0695 * 10^{10}$ **m**   **radius of the Sun**

Chapter XXII

TIME AND SPACE

<u>NEUTRINO CASE No. 20</u>

In Chapter XVI we discussed that time started to exist in the universe after the first oscillators were created. In this chapter we are going to discuss how the concept of space is meaningless without the existence of time. What we mean by meaningless is that space ceases to exist when time stops. To explain this concept we start with the following scenario. In this scenario a friend tells us that he is taking a trip to some remote and unfamiliar part of the world. He will tell us that it takes 8 hours by plane for him to get there. Since we know that modern planes travel at a speed of approximately 600 miles an hour, we can deduce that this god forsaken place is about 5000 miles away. Almost automatically we multiply the speed of the airplane by the expected time length of the trip and we get our answer. But what would happen, when our friend is on the runway with the airplane ready to takeoff, if time stops forever suddenly. The plane will not be able to move even an inch, because no matter how slowly it tries to crawl, it still requires time to move from point A to a hypothetical point B even if point B is only an inch away. The turbojet engine blades will stop rotating because they normally travel continuously from a certain position A to a certain position B in a given time, and this, in a world where time is stopped, is not allowed. Since travel is not allowed in a world without time, we think that the engines will stop, the plane will stop and the passengers will stay forever frozen in the position they were at the moment when time stopped. But our scenario proceeds in a much different and more dramatic manner. The airport runway, just a few moments before covered by a beautiful sunshine, all at once becomes totally dark. The sun rays that reach the Earth in about 8 minutes moving at the speed of light, are stopped because photons cannot travel due to the freezing of time. Now that the sun has turned completely dark, it normally would collapse because of the lack of kinetic energy of all its once vibrating atoms. Its inner layers cannot support the pressure of its outer layers, but collapsing requires movement, and once again this is not allowed. All the protons and neutrons that make up the sun instantaneously lose the photons travelling in their event horizon. The oscillating electromagnetic field of the photons has to disappear. But the energy stored in the electromagnetic fields has to be preserved, and the only way it can be preserved is by converting it into the mass of the neutrinos. Of course, even the plane, the Earth and the rest of the universe will be converted into neutrinos. This event recreates in an instant the scenario described in chapter XIII where all the matter, under the effect of a huge gravitational force, was totally converted into neutrinos. Since the neutrinos are black holes and time inside their event horizon is stopped, we can say that now time has really been stopped all over the universe. The only particles left are neutrinos and neutrinos are singularities and singularities are infinitesimally small. Finally we can say, if what is left is a universe frozen in time, a universe filled only with invisible points, then in this scenario distances are meaningless. This universe itself is a singularity no matter how many infinitesimal small points it contains [1] and space has ceased to exist. In our scenario, by stopping time, we shrank space to an infinitesimally small point.

If we look back to the Universe just before the Big Bang, what we have is just a huge amount of neutrinos in a condition without time and space. It is impossible to differentiate between the Universe before the Big Bang and a present universe where suddenly time stops and therefore space disappears. Of course, what we have described in the previous paragraph is an artificial situation because nothing can stop time suddenly all over the Universe. But slowly, in some parts of the present universe, in places where the gravitational force becomes so strong that not even something travelling at the speed of light can escape, time is being stopped and space becomes a meaningless quantity. We think that the present universe, after many generations where stars and galaxies are created and die, will keep expanding and will continue generating more dark energy and dark matter whenever stars explode into supernovas. Dark matter and dark energy are both caused by the presence of neutrinos in different locations, one in the middle of the visible universe, the other in its outer edges. The amount of dark energy and dark matter, already 95% of the total mass of the universe, will continue to grow because every time a star explodes into a supernova, the amount of neutrinos increases. In the end, only neutrinos will be left in a Universe without time and space. At this point it would be impossible to differentiate between the Universe before the Big Bang and the final stage of the present Universe. We can deduce from this scenario that the life of the Universe, in order to proceed in a cycle, does not have to be expanding from the time of the Big Bang, reach a maximum dimension and start shrinking back to an infinitesimal small entity. In our scenario the Universe keeps on expanding until all the matter is turned into dark matter or dark energy, then, when the only existing particles left are just the neutrinos, time stops, space becomes meaningless and the universe is back to the same conditions as before the Big Bang.

Nobody can tell how long it will take for the present universe to be converted completely into dark matter. Probably the universe will last at least one hundred times the present age. Very likely, the future universe will grow dimmer and dimmer as more dark matter and dark energy keeps on being generated by the exploding supernovas, until it reaches its final stage.

(1)- We know that the number of points is finite otherwise this statement would not be true.

EPILOGUE

Hopefully, after presenting twenty cases of circumstantial evidence, the reader can agree on how it is possible, using the black hole neutrino theory, to explain many of the enigmas of the Universe. This theory first started as a new nuclear particle model, and later was easily expanded to cover many of the mysteries of the Universe. Ten years ago, when we began writing about the neutrino, nobody in the scientific community would share any of these ideas. Today, thanks to discoveries made recently by astrophysicists, a few of these ideas have become widely accepted.

In the last few years astrophysicists have been able to search deep into the Universe using newly built observatories on Earth and space. From these exciting observations, they were able to formulate new theories and some are pertinent to our neutrino case. Among these new ideas, we'd like to bring out a few that we now share with most of these scientists.

- Dark matter is made of neutrinos. From the beginning, this has been a basic concept in our theory as it is presented in Chapter XI.
- The Universe before the Big Bang was much smaller than an atom. This idea was introduced in our previous books and is repeated in Chapter XIII.
- Neutrinos were the first particles after the Big Bang. Our book not only agrees with this statement but believes that neutrinos were the only particles even before the Big Bang.

- The pictures of the Bullet Galaxy show that under certain conditions the dark matter trails behind the visible matter. This was one of the original predictions of our theory.

We are pleased that a small part of our theory is now shared by the most progressive astrophysicists of our time. Unfortunately, no progress has been made on the nuclear particles part of our theory. The main body of our theory states that neutrinos are black holes and, of course, this part has not yet been agreed upon by any scientist. For this event to happen, it would require nuclear particles physicists to make dramatic changes to the Standard Model. These are some of the statements in our book that would make these changes necessary.

- The neutrino cannot be classified as a lepton because no magnetic moment has ever been found. The neutrino is a unique particle that behaves as a singularity.
- Quarks are a good mathematical tool to explain what happens inside high energy particle accelerators.
- The neutrino has no antiparticle. After the Big Bang the neutrino, for the first few years, was the only existing particle; at this time, the existence of an antiparticle neutrino would have jeopardized the development of the Universe.
- It is impossible to create a stable particle that has a fraction of an electron charge.
- The magnetic moment of a charged particle is inversely proportional to its mass.

- The neutrino appears in the interactions of all stable nuclear particles, because it is at the center of every stable particle.
- A revised Standard Model needs to address the gravitational force because it is one of the four basic forces of physics.
- All stable particles have the same spin number because the spin is generated by the neutrino in their center.

These changes are too comprehensive to expect to be incorporated in the near future. If, looking at the past, we were to make any prediction for the future, we should realize that nothing much has changed in the Standard Model in the last forty years. Twenty years ago, what was considered the major discovery of the W and Z bosons is now in doubt and these bosons have been renamed as pseudo particles. Currently, most of the experiments in nuclear physics take place in the many particle accelerators facilities throughout the World. There, while mostly similar types of experiments are repeated over and over again, physicists continue looking for different results. They continue searching for the elusive Higgs boson. Einstein once defined this kind of behavior for a scientist as insanity. More hope is provided by the considerable new efforts being made on separate projects aimed to uncover the true nature of the neutrino. As we explained in the Introduction, neutrinos interact so little with the surrounding matter that data is being gathered very slowly from these sites. Therefore, with much regret, we do not expect any major changes in the Standard Model to happen for at least another twenty long and monotonous years.

The Neutrino Case

The Neutrino Case

Appendix   A

Magnetic Moment Computations

Whenever there is an electrical charge or current rotating in a circular orbit, a magnetic field is generated. The magnetic field, in this case, is more properly called a magnetic dipole moment as shown in *Figure A.1* This dipole acts exactly as a small magnet and aligns itself with any exterior magnetic field just as the needle of a compass aligns with the Earth's magnetic field. The magnitude of the dipole is computed by multiplying the current I by the area A:

Dipole Moment = M = I x A *(Equation A.1)*

In the case of the charge rotating in a circle, we compute the equivalent current I by multiplying the given charge q by the frequency v at which it revolves around the circle:

Dipole Moment = M = I x A= qv x A

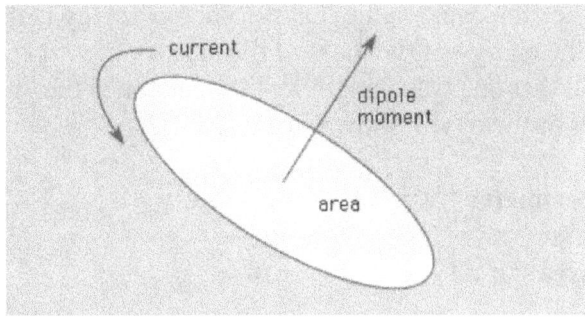

Figure A.1-Magnetic Dipole Moment

Exactly the same magnetic field effect happens when we have a photon rotating around a neutrino or singularity since, as we explained in Chapter I, this condition generates a rotating charge and a rotating charge creates a current and, therefore, a magnetic dipole.

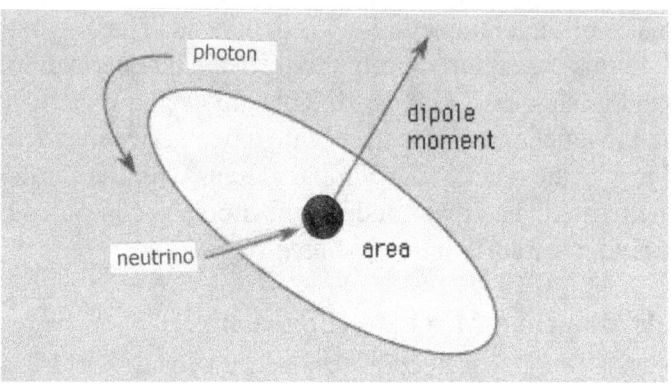

Figure A.2- The Electron Dipole

Figure A.2 depicts the dipole of an electron. In order to compute the intensity of the magnetic field of the electron, we have to find the area and the current as in *Equation A.1*. To find the area, we assume that the photon energy is the same as the energy of the mass of the electron, or .511 MeV, and that one wavelength λ of the photon is the perimeter of the dipole shown in *Figure A.2*:

perimeter = λ = 2 π r

area = π r²

To find λ we equate the energy of the photon to the energy of the electron:

$$h\nu_{photon} = E_{electron}$$

From this equation we can compute the frequency of the photon:

$$\nu = E_{electron} / h = 1.235 \times 10^{20} \text{ Hz}$$

And from the frequency we compute the wavelength:

$$\lambda = c / \nu = 2.428 \times 10^{-12} \text{ m}$$

Then we equate one wavelength to the perimeter of the dipole:

$$\lambda = 2 \pi r ; r = \lambda / 2 \pi$$

Therefore, we compute the area of the dipole:

$$A = \pi r^2 = \pi (\lambda / 2 \pi)^2 = \lambda^2/4 \pi = .0469 \times 10^{-23} \text{ m}^2$$

We compute the current **I** that produces the magnetic moment by multiplying the charge of the electron **q** by the frequency of the photon **v**:

$$I = q \nu = 1.601 \times 10^{-19} \times 1.235 \; 10^{20} = 19.76$$
(Amps)

And, finally, we compute the magnetic moment by multiplying the area by the current:

[147]

$$M = I \times A = 927.3 \times 10^{-26} \ (J \ T^{-1})$$

The figure that we computed is only one part in one thousand smaller than the actual magnetic field of the electron. The difference can be explained by the fact that, in the equations, we did not take into consideration the mass of the neutrino; later on, using more accurate calculations, our model will show the mass of the neutrino to be 600 ev. This mass is within the limits of the oscillating neutrino theory.

In our small black hole model, any time a photon is captured by a singularity an anomalous electromagnetic wave is generated; we call it anomalous because the inner wave, being smaller than the outer wave, creates a field imbalance. The net result of this imbalance is that, in the particle, there is the equivalent of a single charge rotating at the speed of light around the singularity. This charge can be positive or negative and its intensity is always the equivalent of the charge of the electron. The same effect happens in all the nuclear particles with the exception of the neutrino. The reason that the neutrino does not have a magnetic field is easily explainable by our model. Our model shows the neutrino in the center of the particle, and whenever there is no photon associated with it, there is no rotating charge and therefore no magnetic field.

At this time, nuclear physicists have not found any magnetic field associated with the neutrino, even after repeated experiments with the most accurate and sensitive instrumentation.
Now we need to introduce a new nuclear particle that is not part of the stable particles family. This particle is the muon.

Muons are created when cosmic rays collide with the Earth's upper atmosphere. Their life span is about 2.2 μsec. In the following table, we compare certain characteristics of the muon and the electron.

Particle	Mass(Mev)	Magnetic Moment(JT-ɨ)
Electron	0.511	-928.48
Muon	105.66	-4.490

Magnetic Moment(JT^{-1}) 10 $^{-26}$

Referencing this table, if we divide the mass of the muon by the mass of the electron, we have a ratio equal to 206.8. And if we divide the magnetic moment of the muon by the magnetic moment of the electron muon, we have a ratio equal 1 / 206.8. From this result we can deduce that the magnetic moments in these two particles are inversely proportional to the masses of the particles. The greater the particle mass, the smaller the magnetic moment. This time, in computing the muon magnetic moment, we find the magnetic moment general formula. We equate, first, the energy of the photon to the energy of the muon:

$$h\nu_{photon} = E_{\ muon} = m_{muon}\ C^2$$

$$\nu = E_{muon}\ /\ h = m_{muon}\ C^2\ /\ h$$

Considering the perimeter equal to one wave length λ, we find the area of the magnetic field.

$$A = \pi\ r^2 = \pi\ (\ \lambda\ /2\ \pi\ )^2 = \lambda^2\ /\ 4\ \pi = (\ c\ /\ \nu\ )^2/\ 4\ \pi$$

$$M = I\ x\ A = q\ \nu\ (\ c\ /\ \nu\ )^2/4\ \pi = q\ c^2/4\ \pi\ \nu$$

$$M = \underline{q\ h\ /4\ \pi\ m}$$

The above formula is the general formula for the magnetic moment and was derived using our model. This formula clearly indicates that the magnetic moment is inversely proportional to the mass of the particle.

When we apply this formula $M = q\ h\ /4\ \pi\ m$ to the muon, we compute a magnetic moment that is only a few thousandths off the actual value.

With our model we have been able to solve the enigma that smaller particles have a stronger magnetic field than the larger particles; furthermore, our model has provided the means for computing the magnetic dipole associated with these particles with great accuracy. In general, the electron and the muon are classified in the family of the leptons, while the proton is classified in the family of the baryons.

Appendix B

The Standard Model

This appendix is provided to give the reader a direct comparison between the Standard Model and our New Model Theory.

The Standard Model is a theory which describes three of the four known <u>fundamental interactions</u> between the <u>elementary particles</u> that make up all <u>matter</u>. It is a <u>quantum field theory</u> developed between <u>1970</u> and <u>1973</u> which is consistent with both <u>quantum mechanics</u> and <u>special relativity</u>. To date, almost all experimental tests of the three forces described by the Standard Model have agreed with its predictions. However, the Standard Model falls short of being a <u>complete theory of fundamental interaction</u>, primarily because of its lack of inclusion of <u>gravity</u>, the fourth known fundamental force.

It is understood that the entire dynamics of the Universe can be explained in terms of matter and by the forces that act on it. The Standard Model is divided in a similar manner: ordinary matter particles (**baryons** and **leptons**), force mediating particles (**bosons**), and the <u>Higgs</u> particle (also a boson).

Technically, quantum field theory provides the mathematical framework for the Standard Model. Consequently, each type of particle is described in terms of a mathematical field. For a technical description of the fields and their interactions, see Standard model (basic details).

The matter particles described by the Standard Model all have an intrinsic spin whose value is determined to be 1/2, making them fermions. For this reason, they follow the Pauli Exclusion Principle. Apart from their antiparticle partners, a total of twelve different matter particles are known as of early 2007. Six of these are classified as quarks (up, down, strange, charm, top and bottom), and the other six as leptons (electron, muon, tau, and their corresponding neutrinos).

These particles carry charges which make them susceptible to the fundamental forces (described in the next subsection).

- Each quark carries any one of three color charges – red, green or blue, enabling them to participate in strong interactions.

- The up-type quarks (up, charm, and top quarks) carry an electric charge of +2/3, and the down-type quarks (down, strange, and bottom) carry an electric charge of –1/3, enabling both types to participate in electromagnetic interaction.

- Leptons do not carry any color charge – they are color neutral, preventing them from participating in strong interactions .The down-type leptons (the electron, the muon, and the tau lepton) carry an electric charge of –1, enabling them to participate in electromagnetic interactions.

- The up–type leptons (the neutrinos) carry no electric charge, preventing them from participating in electromagnetic interactions.
- Both, quarks and leptons carry a handful of flavor charges (see flavor), enabling all particles to interact via the weak nuclear interaction.
- Pairs from each group (one up-type quark, one down-type quarks, a lepton and its corresponding neutrino) form a generation. Corresponding particles within each generation are identical to each other apart from their masses and flavors.

The force mediating particles described by the Standard Model all have an intrinsic spin whose value is 1, making them bosons. As a result, they do not follow the Pauli Exclusion Principle. The different types of force mediating particles are described below.

- The photons mediate the familiar electromagnetic force between electrically charged particles (these are the quarks, electrons, muons, tau, W^+ and W^-). They are massless and are described by the theory of quantum electrodynamics.
- The W^+, W^-, and Z^0 gauge bosons mediate the weak nuclear interactions between particles of different flavors (all quarks and leptons). They are massive, with the Z^0 being more massive than the equally massive W^+ and W^-. An interesting feature of the weak force is that interactions involving the W^+ and W^- gauge bosons act on exclusively *left-handed* particles (those particles whose spins - 1/2h). The *right-handed* particles are completely neutral to the W bosons. Furthermore, the W^+ and W^- bosons carry an

electric charge of +1 and −1 making those susceptible to electromagnetic interactions. The electrically neutral Z^0 boson acts on particles of both spins, but preferentially on *left-handed* ones. The weak nuclear interaction is unique in that it is the only one that selectively acts on particles of different spins; the photons of electromagnetism and the gluons of the strong force act on particles without such prejudice. These three gauge bosons along with the photons are grouped together which collectively mediate the electroweak interactions.

The Twenty Neutrino Cases

Glossary

ANTIMATTER: In particle physics, **antimatter** is the extension of the concept of the antiparticle to matter, where antimatter is composed of antiparticles in the same way that normal matter is composed of particles. For example, an antielectron (a positron, an electron with a positive charge) and an antiproton (a proton with a negative charge) could form an antihydrogen atom in the same way that an electron and a proton form a *normal matter* hydrogen atom. Furthermore, mixing matter and antimatter would lead to the annihilation of both in the same way that mixing antiparticles and particles does, thus giving rise to high-energy photons (gamma rays) or other particle-antiparticle pairs.

ASTROPHYSIC: The branch of astronomy that deals with the physics of the universe, including the physical properties (luminosity, density, temperature, and chemical composition) of celestial objects such as galaxies, stars, planets, and the interstellar medium, as well as their interactions.

BARYONS: See Appendix B

BIG BANG: The **Big Bang** is the cosmological model of the initial conditions and subsequent development of the universe supported by the most comprehensive and accurate explanations from current scientific evidence and observation. As used by cosmologists, the term *Big Bang* generally refers to the idea that the universe has expanded from a primordial hot and dense initial condition at some finite time in the past, and continues to expand to this day.

BLACK HOLES: A **black hole** is a region of space in which the gravitational field is so powerful that nothing, including electromagnetic radiation (e.g. visible light), can escape its pull after having fallen within its event horizon. The term derives from the fact that absorption of visible light renders the hole's interior invisible, and indistinguishable from the black space around it.

BOSONS: See Appendix B

DARK MATTER: In astronomy and cosmology, **dark matter** is hypothetical matter that is undetectable by its emitted radiation, but whose presence can be inferred from gravitational effects on visible matter. Dark matter is postulated to explain the flat rotation curves of spiral galaxies and other evidence of "missing mass" in the universe.
According to present observations of structures larger than galaxies, as well as Big Bang cosmology, dark matter and dark energy account for the vast majority of the mass in the observable universe. Dark matter also plays a central role in the structure formation and galaxy evolution, and has measurable effects on the anisotropy of the cosmic microwave background. All these lines of evidence suggest that galaxies, clusters of galaxies, and the universe as a whole contain far more matter than that which interacts with electromagnetic radiation.

ELECTRON: The **electron** is a subatomic particle that carries a negative charge. Electrons participate in gravitational, electromagnetic and weak interactions. Like its rest mass and elementary charge, the intrinsic angular momentum (or spin) of an electron has a constant value. In the collision of an electron and a positron, the electron's antiparticle, both are annihilated. An electron-positron pair can be produced from gamma ray photons with sufficient energy.

<u>EVENT HORIZON</u>: In general relativity, an **event horizon** is a boundary in spacetime, most often an area surrounding a black hole. Light emitted from within the horizon can never reach an outside observer.

<u>GENERAL THEORY OF RELATIVITY</u>: The **general theory of relativity** is the geometric theory of gravitation published by Albert Einstein in 1916. It is the current description of gravity in modern physics. It unifies special relativity and Newton's law of the universal gravitation, and describes gravity as a property of the geometry of space and time, or spacetime. In particular, the curvature of spacetime is directly related to the four-momentum (mass-energy and linear momentum) of whatever matter and radiation are present. The relation is specified by the Einstein field equations, a system of partial differential equations.

<u>GRAVITATIONAL POTENTIAL ENERGY</u>:
Gravitational potential energy is the energy an object possesses because of its position in a gravitational field. The most common use of gravitational potential energy is for an object near the surface of the Earth where the gravitational acceleration can be assumed to be constant at about 9.8m/s^2. Since the zero of gravitational potential

energy can be chosen at any point (like the choice of the zero of a coordinate system), the potential energy at a height h above that point is equal to the work which would be required to lift the object to that height with no net change in kinetic energy. Since the force required to lift it is equal to its weight, it follows that the gravitational potential energy is equal to its weight times the height to which it is lifted.

KINETIC ENERGY: The **kinetic energy** of an object is the extra energy which it possesses due to its motion. It is defined as the work needed to accelerate a body of a given mass from rest to its current velocity. Having gained this energy during its acceleration, the body maintains this kinetic energy unless its speed changes. Negative work of the same magnitude would be required to return the body to a state of rest from that velocity.

LEPTONS: Leptons are a family of elementary particles, alongside quarks and gauge bosons (also known as force carriers). Like quarks, leptons are fermions (spin-1/2 particles) and are subject to the electromagnetic force, the gravitational force, and weak interaction. But unlike quarks, leptons do not participate in the strong interaction. Leptons are an important part of the Standard Model, especially the electrons which are one of the components of atoms, alongside protons and neutrons. Exotic atoms with muons and tauons instead of electrons can also be synthesized.

NEUTRINOS: Elementary particles that travel to the speed of light; lack an electric charge; are able to pass through ordinary matter almost undisturbed and are thus extremely difficult to detect. Neutrinos have a minuscule, but nonzero mass. They are usually denoted by the Greek letter v (nu).

NEUTRON: The **neutron** is a subatomic particle with no net electric charge and a mass slightly larger than that of a proton. Neutrons are usually found in atomic nuclei. The nuclei of most atoms consist of protons and neutrons, which are therefore collectively referred to as nucleons. The number of protons in a nucleus is the atomic number and defines the type of element the atom forms. The number of neutrons determines the isotope of an element. For example, the carbon-12 isotope has 6 protons and 6 neutrons, while the carbon-14 isotope has 6 protons and 8 neutrons.

NUCLEON: The collective name for a proton or a neutron. These subatomic particles are the principal constituents of atomic nuclei and therefore of most matter in the universe. The proton and neutron share many characteristics.They have the same intrinsic spin, nearly the same mass, and similar interactions with other subatomic particles, and they can transform into one another by means of the weak interactions. Hence it is often useful to view them as two different states or configurations of the same particle, the nucleon. Nucleons are small compared to atomic dimensions and relatively heavy.

Their characteristic size is of order 1/10,000 the size of a typical atom, and their mass is of order 2,000 times the mass of the electron.

NUCLEUS: The **nucleus** of an atom is the very dense region, consisting of nucleons (protons and neutrons), at the center of an atom. Almost all of the mass in an atom is made up from the protons and neutrons in the nucleus with a very small contribution from the orbiting electrons. The diameter of the nucleus is in the range of 1.6 fm (1.6 x 10^{-15}) (for a proton in light hydrogen) to about 15 fm (for the heaviest atoms, such as uranium). These dimensions are much smaller than the size of the atom itself by a factor of about 23,000 (uranium) to about 145,000 (hydrogen).

PAULI EXCLUSION PRINCIPAL: in 1925, Wolfgang Pauli gave physics his exclusion principle as a way to explain the arrangement of electrons in an atom. His hypothesis was that only one electron can occupy a given quantum state. That is, each electron in an atom has a unique set of quantum numbers (the principle quantum number which gives its energy level, the magnetic quantum number which gives the direction of orbital angular momentum, and the spin quantum number which gives the direction of its spin). If this principle did not hold, all of the electrons in an atom would pile up in the lowest energy state (the K shell). In fact, we now know that the Pauli Exclusion Principle holds for not just electrons but for any fermions (half-integer spin particles like electrons, protons, neutrons, muons, and many more.

PARTICLE (SUBATOMIC): A **subatomic particle** is an elementary or composite particle smaller than an atom. Particle physics and nuclear physics are concerned with the study of these particles, their interactions, and non-atomic matter. Subatomic particles include the atomic constituents electrons, protons, and neutrons.

PARTICLE SPIN: As the name indicates, the spin has originally been thought of as a rotation of particles around their own axis. This picture is correct insofar as spins obey the same mathematical laws as do quantized angular momenta. On the other hand, spins have some peculiar properties that distinguish them from orbital angular momenta: spins may have half-integer quantum numbers, and the spin of charged particles is associated with a magnetic dipole moment in a way (g-factor different from 1) that is incompatible with classical physics. By spin 1 or 0.5, we really mean 1 * hbar or 0.5 * hbar. Whether the particle has integer spin or half-integer spin is very influential on it's behavior. Integer spin particles (e.g. photons) are called bosons, follow the laws known as Bose-Einstein statistics, while half integer spins are called Fermions (e.g. electrons) and follow Fermi-Dirac statistics.

PION: Pion or pi meson, lightest of the meson family of elementary particles. The existence of the pion was predicted in 1935 by Hideki Yukawa, who theorized that it was responsible for the force of the strong interactions holding the atomic nucleus together.

PHOTON: In physics, the **photon** is an elementary particle, the quantum of the electromagnetic field and the basic unit of light and all other forms of electromagnetic radiation. It is also the force carrier for the electromagnetic force. This force's easily visible human-scale effects and applications, from sunlight to radiotelephones, are because the photon has no mass and thus can produce interactions at long distances. Like all elementary particles, the photon is governed by quantum mechanics and so exhibits wave-particle duality: that is, it exhibits both wave and particle properties. For example, a single photon may undergo refraction by a lens or exhibit wave interference, but also act as a particle giving a definite result when its location is measured.

PROTON: The **proton** is a subatomic particle with an electric charge of +1 elementary charge. It is found in the nucleus of each atom but is also stable by itself and has a second identity as the hydrogen ion, H+.

QUARKS: See Appendix B

RUTHERFORD: Ernest Rutherford, (30 August 1871 - 19 October 1937) was a New Zealand chemist who became known as the father of nuclear physics. He discovered that atoms have a small charged nucleus, and thereby pioneered the Rutherford model (or planetary model, which later evolved into the Bohr model or orbital model) of the atom, through his discovery of Rutherford scattering with his gold foil experiment. He was awarded the Nobel Prize in Chemistry in 1908.

SINGULARITY: A **gravitational singularity** (sometimes **spacetime singularity**) is, approximately, a place where quantities which are used to measure the gravitational field become infinite. Such quantities include the curvature of spacetime or the density of matter. More accurately, a spacetime with a singularity contains geodesics which cannot be completed in a smooth matter. The limit of such a geodesic is the singularity.

SCHWARZSCHILD: The relativistic Schwarzschild model of a static spherically symmetric star of uniform density is extended to the case in which the star is allowed to expand or contract. It is shown that the velocity of expansion or contraction of such a star obeys a law of the form v=H r, where r is the distance from the center. It is further shown that all homologous perturbations of such a star lead either to infinite expansion or to gravitational collapse.

STANDARD MODEL: See Appendix B

SUPERNOVA: (plural: supernovae) is a stellar explosion. Supernovae are extremely luminous and cause a burst of radiation that often briefly outshines an entire galaxy, before fading from view over several weeks or months. During this short interval, a supernova can radiate as much energy as the Sun could emit over its life span.

The explosion expels much of all of a star's material at a velocity of up to a tenth the speed of light, driving a shock wave into the surrounding interstellar medium. This shock wave sweeps up an expanding shell of gas and dust called a supernova remnant.

WORMHOLE: In physics, a **wormhole** is a hypothetical topological feature of spacetime that is fundamentally a "shortcut" through space and time. Spacetime can be viewed as a 2D surface, and when "folded" over, a wormhole bridge can be formed. A wormhole has at least two mouths which are connected to a single throat or tube. If the wormhole is **traversable**, matter can "travel" from one mouth to the other by passing through the throat. While there is no observable evidence for wormholes, spacetimes-containing wormholes are known to be valid solutions in general relativity.

Fundamental Constants of Physics

Speed of light in vacuum, C   299792458 m s^{-1}

Electric constant, ϵ 8.854187817 x 10^{-12} F m^{-1}

Planck constant, h   6.62606876 x 10^{-34} J s

Electron charge, e   1.602176642 x 10^{-19} C

Electron equivalent energy   0.510998902 Mev

Electron mass, m$_e$ 9.10938188 kg x 10^{-31}

Electron magnetic moment, μ_e -928.476362 x 10^{-26}JT^{-1}

Muon equivalent energy   105.6583568 Mev

Muon magnetic moment, μ_μ -4.49044813 x 10^{-26}T^{-1}

Proton mass, m$_p$ 1.67262158 x 10^{-27} kg

Proton equivalent energy   938.271998 Mev

Proton magnetic moment, μ_p 1.410606633 x 10^{-26} JT^{-1}

Neutron equivalent energy   939.565330 Mev

Neutron magnetic moment, μ_n-0.96623640 x 10^{-26} J T^{-1}

Gravitational constant, G   6.67 x 10^{-11} n m^2/ kg^2

References

Robert K. Adair- The Great Design (1987)

V.D. Barger- Classical Electricity and Magnetism (1987)

John M. Blatt – Theoretical Nuclear Physics (1952)

Sidney Borowits- Essential Of Physics (1967)

K.G. Budden- The Propagation of Radio Waves (1988)

J.Z. Buckwald- The Rse of the Wave Theory of Light (1989)
L.Eyges- The Classical Electromagnetic Field (1980)

Richard P. Feynman- Lectures On Gravitation (1995)

Richard P. Feynman- QED (1985)

Robert L. Forward- Mirror Matter (1988)

Brian Greene- The Elegant Universe (1999)

Heald Marion- Classical Electromagnetic Radiation (1994)

Eugen Merzbacher- Quantum Mechanics (1998)

M.H.Nayfeh- Electricity and Magnetism (1985)

W. Pauli- Theory of Relativity (1981)

Donald H. Perkins- Introduction To High Energy
Physics (2000)

W.G.V. Rosser Classical Electromagnetism via Relativity
(1968)
O. Svelto- Principles of Lasers (1989)

The Neutrino Case

The Neutrino Case

The Neutrino Case